Aerosol–Cloud–Climate Interactions

This is Volume 54 in the
INTERNATIONAL GEOPHYSICS SERIES
A series of monographs and textbooks
Edited by RENATA DMOWSKA and JAMES R. HOLTON

A complete list of books in this series appears at the end of this volume.

Aerosol–Cloud–Climate Interactions

Edited by

Peter V. Hobbs
DEPARTMENT OF ATMOSPHERIC SCIENCES
UNIVERSITY OF WASHINGTON
SEATTLE, WASHINGTON

ACADEMIC PRESS, INC.
Harcourt Brace & Company
San Diego   New York   Boston
London   Sydney   Tokyo   Toronto

Front cover photograph: Satellite (NOAA 10) picture in near infrared (Channel 2) from passage 27 August 1988 around 0845 UTC. See Figure 7.2. (Satellite Receiving Station, Dundee University, U. K. From Raustein et al., 1991.)

This book is printed on acid-free paper. ⊚

Academic Press, Inc.
1250 Sixth Avenue, San Diego, California 92101-4311

United Kingdom Edition published by
Academic Press Limited
24–28 Oval Road, London NW1 7DX

Library of Congress Cataloging-in-Publication Data

Aerosol–cloud–climate interactions. / edited by Peter V. Hobbs.
   p.   cm. – (International geophysics ; v. 54)
   Includes index.
   ISBN 0-12-350725-1
   1. Aerosols.   2. Cloud physics.   3. Climatology   I. Hobbs, Peter
   Victor, 1936-   II. Series.
   QC882.42.A34   1993
   551.5–dc20   92-41627
                                                                                    CIP

PRINTED IN THE UNITED STATES OF AMERICA
93   94   95   96   97   98   E B   9   8   7   6   5   4   3   2   1

Contents

Chapter 1 Tropospheric Aerosols
Ruprecht Jaenicke

Chapter 2 Aerosol–Cloud Interactions
Peter V. Hobbs

Chapter 3 Aerosol–Climate Interactions
Harshvardhan

Chapter 4   Microphysical Structures of Stratiform and Cirrus Clouds

Andrew J. Heymsfield

Chapter 5   Radiative Properties of Clouds

Michael D. King

Chapter 6   Radiative Effects of Clouds on Earth's Climate

D. L. Hartmann

Chapter 7   Parameterization of Clouds in Large-Scale Numerical Models

Hilding Sundqvist

Chapter 8   Stratospheric Aerosols and Clouds

M. Patrick McCormick, Pi-Huan Wang, and Lamont R. Poole

Contributors

Numbers in parentheses indicate the pages on which the authors' contributions begin.

HARSHVARDHAN (75), Department of Earth and Atmospheric Sciences, Purdue University, West Lafayette, Indiana 47907

D. L. HARTMANN (151), Department of Atmospheric Sciences, University of Washington, Seattle, Washington 98195

ANDREW J. HEYMSFIELD (97), National Center for Atmospheric Research, Boulder, Colorado 80307

PETER V. HOBBS (33), Department of Atmospheric Sciences, University of Washington, Seattle, Washington 98195

RUPRECHT JAENICKE (1), Institute of Physics of the Atmosphere, University Mainz, 6500 Mainz, Germany

MICHAEL D. KING (123), NASA Goddard Space Flight Center, Greenbelt, Maryland 20771

M. PATRICK McCORMICK (205), Atmospheric Sciences Division, NASA Langley Research Center, Hampton, Virginia 23681

LAMONT R. POOLE (205), Atmospheric Sciences Division, NASA Langley Research Center, Hampton, Virginia 23681

HILDING SUNDQVIST (175), Department of Meteorology, Stockholm University, S-106 91 Stockholm, Sweden

PI-HUAN WANG (205), Science and Technology Corporation, Hampton, Virginia 23666

Preface

Together with molecular scattering from gases, aerosol and clouds determine what fraction of the solar radiation incident at the top of the atmosphere reaches the earth's surface, and what fraction of the longwave radiation from the earth escapes to space. Consequently, aerosol and clouds play an important role in determining the earth's climate. Much remains to be learned about the properties of atmospheric aerosol and clouds and their effects on the radiative balance of the earth. The problem stems, in part at least, from the fact that unlike gases, aerosol and clouds are distributed very unevenly in the atmosphere. Also, many of the relevant properties of aerosol and clouds are not well defined. Finally, there are strong interactions between aerosol, clouds, and climate (see the accompanying figure), many of which we are only just beginning to understand, let alone incorporate into climate models.

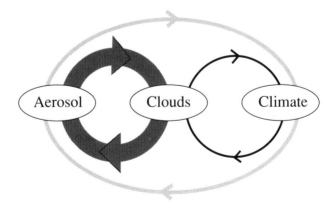

AEROSOL - CLOUD - CLIMATE INTERACTIONS

This book provides an overview of our present understanding of atmospheric aerosol and clouds, of the six potential interactions indicated in the figure, and of the outstanding problems that remain to be solved if we are to understand this complex system in both its natural and (anthropogenically) perturbed state. The first chapter reviews the sources, sinks, and properties of atmospheric aerosol. In the second chapter the effects of aerosol on clouds, and the effect of clouds on aerosol, are described. Chapter 3 is concerned with the direct effects of aerosol on the radiation balance of the earth and climate. A review of the principal forms of clouds, and their microstructures, is presented in Chapter 4. This is followed, in Chapters 5 and 6, by reviews of the radiative properties of clouds and the current understanding of the effects of clouds on the radiative balance of the earth. Chapter 7 provides an account of recent attempts to include fairly detailed information on clouds in operational numerical prediction models and general circulation models. The book concludes with a discussion of stratospheric aerosol and clouds.

The framework for this book was conceived at the XX Assembly of the International Union of Geodesy and Geophysics, Vienna, Austria, 13–20 August 1991. I am grateful to the authors for their willingness to expand considerably on the scope of their original presentations in order to provide the comprehensive description of the subject that is contained in this book.

Peter V. Hobbs

Chapter 1 | Tropospheric Aerosols

Ruprecht Jaenicke
Institute for Physics of the Atmosphere
University Mainz

Tropospheric aerosols derive from many sources. They are distributed in the atmosphere through turbulence and direct atmospheric transport of air masses. They are removed by precipitation, coagulation, and sedimentation (dry impaction). One of the sources is nonprecipitating clouds. Clouds serve not only as chemical reactors, but also as particle sources; this source of 3000 Tg yr^{-1} of aerosol mass is comparable in strength to other major sources like oceans and deserts. This chapter presents the most recent estimates on aerosol sources and results on aerosol distributions and on aerosol parameter data.

I. Introduction

Public concern with global and climate change might encourage a yearning for an atmosphere without trace substances, except possibly stratospheric ozone. However, if atmospheric aerosols[1] were not present, cloud and precipitation distributions would be quite different and we would live on an earth with a constantly burning sun and quite a different climate. Fortunately, aerosols are omnipresent in our atmosphere. Yet, much remains to be learned about this generally invisible component of the air.

In this chapter, we review briefly some important aspects of tropospheric aerosols, including their sources, geographic distributions, sizes, chemical properties, and their transport and residence times. However, it is not the aim of this chapter to provide a comprehensive survey of the subject. Instead, we will emphasize the physical and chemical properties of tropospheric aerosols. This information will provide a backdrop to the reviews of other aspects of atmospheric aerosols, clouds, and aerosol–cloud–climate interactions given elsewhere in this book.

---

[1]The term *aerosols* in addition to *aerosol* is used to indicate the presence of types of different bodies comparable to air masses or clouds.

Aerosol–Cloud–Climate Interactions   1

II.   Historical Understanding of Tropospheric Aerosols

Without defining it, early observations of atmospheric aerosols were carried out by visual observations (Christ, 1783). Aitken (1888) designed an instrument to measure what we now call Aitken particles (he called it *dust*). Mie (1908) developed a theoretical approach for handling optically active particles, today often called Mie particles. Electrically charged atmospheric particles (large ions) tended to be treated separately (Nolan and Doherty, 1950), as were radioactive particles (Chamberlain and Dyson, 1956). Particles that serve to nucleate cloud drops (cloud condensation nuclei) and ice in clouds (ice nuclei) also have been generally considered separately (Twomey, 1959). Until recently, biological aerosols (other than pollen and spores) were almost neglected by the atmospheric aerosol community (Orsini et al., 1986; d'Almeida et al., 1991). Even today, many conferences, symposia, and monographs treat only portions of the total atmospheric aerosol.

This type of departmentalization of atmospheric particles originated, in part, from the use of special measuring techniques. Even when Junge (1952*a*) published the first view of a continuous aerosol size distribution, covering radii from 0.01 to 10 μm, the picture of different aerosol populations was still strong. Only subsequently did the view of a continuous atmospheric aerosol size distribution, with an analytical power function to describe it, shine through (Junge, 1952*b*). The impact Junge made with his model of an atmospheric aerosol size distribution was so overwhelming that people even talked about a "law" (e.g., Monahan and Muircheartaigh, 1980; d'Almeida et al., 1991), rather than a descriptive function.

The concept of different aerosol populations also influenced research on the chemistry of atmospheric aerosols. Thus, there were publications on radioactive particles (Chagnon and Junge, 1965), sulfate (Wagman et al., 1967; d'Almeida et al., 1991), nitrate (Savoie and Prospero, 1982), sea salt (Woodcock and Gifford, 1949), organic particles (Grosjean et al., 1982), and pollen (Cour and van Campo, 1980), to name a few. This occurred despite the fact that the concept of "mixed nuclei" (i.e., different compounds mixed together to form a single particle) had long been understood (Junge, 1952*a*).

The subdivision of aerosols into different populations might also reflect a desire to classify things to better understand them. Thus, after Junge introduced the continuous atmospheric aerosol size distribution, he also classified them geographically into maritime, continental, and background aerosols (Junge, 1963), and by size into Aitken (0.001–0.1 μm radius),[2] large (0.1–1 μm), and giant (>1 μm) particles. Whitby (1973) pointed out that the plotting of aerosol size distributions on log–log plots, together with the use of an analytical power function, blurred the distinctive pattern of atmospheric aerosol modes. He introduced the terms *nucleation mode* (0.001–0.1 μm),[3] *accumulation mode* (0.1–1 μm),

---

[2] In this chapter we will use radius for characterizing particle size. However, authors often do not distinguish between radius and diameter because of the uncertainties in aerosol size measurements.

and *coarse particle mode* (>1 μm). While Junge had emphasized measuring method and particle size, Whitby's classification focused on production mechanism and particle size. The nucleation mode was produced by gas-to-particle conversion (GPC), the accumulation mode by coagulation and heterogeneous condensation, and the coarse mode by mechanical processes. While these processes might act rapidly enough in regions of very high aerosol concentration typical of urban areas, they may not be as important for the tropospheric aerosols in general (e.g., sea salt aerosol: Blanchard and Woodcock, 1980; mineral aerosol: d'Almeida and Schütz, 1983). Whitby's ideas hold sway today in describing atmospheric aerosol size distributions by the sum of three lognormal functions (see Table 2).

III. Contemporary Understanding of Tropospheric Aerosols

A. Particle Sources and Strengths

There are two major sources of atmospheric aerosols: widespread surface sources and spatial sources. By widespread surface sources we mean sources at the base of the atmospheric volume (e.g., oceans and deserts). Spatial sources refer to those within the atmospheric volume (e.g., GPC and clouds). Additional point sources, such as volcanoes, are globally important in their influence on the stratosphere. Otherwise, due to short tropospheric residence times, aerosols from point sources affect mainly regional and local scales (as demonstrated by the Chernobyl disaster of 1986 and the Kuwait oil fires of 1991—Hobbs and Radke, 1992). Extraterrestrial sources are negligible.

1. Widespread Surface Sources

Our general understanding of surface sources has not changed appreciably in recent years, but estimates of their strengths have been modified.

a. Biogenic Sources
Biogenic and organic particulate matter must be differentiated. Biogenic matter is a primary product (i.e., particles are injected directly into the atmosphere, either from the biosphere or from other surfaces—resuspension?—see below); organic particles are secondary products (i.e., particles that form in the atmosphere—see Section A.2).

Pollen, spores, and fragments of animals and plants are usually in the micrometer size range, while bacteria, algae, protozoa, fungi, and viruses are smaller. Up to now, mostly the viable (i.e., those that live and reproduce) species have been investigated. They have been identified because of their culture-forming ability. Dead biogenic material has hardly been investigated at all.

---

[3] In the original literature given as diameter.

Certain biological particles (living or dead) may act as ice nuclei in clouds (e.g., Levin and Yankofsky, 1988). This is the case for bacteria (Maki and Willoughby, 1978; Yankofski et al., 1981; Schnell and Vali, 1976; Vali et al., 1976; Levin et al., 1980; Hirano et al., 1978; Lindemann et al., 1982), pollen (Dingle, 1966), and leaf abrasions (Schnell, 1982).

Microorganisms and other biogenic particles originate from both natural and anthropogenic sources. Their existence in the atmosphere is temporary. The atmosphere is only a vehicle for the transport of these particles; it is not a habitat where they feed and reproduce.

The oceans are a prominent source of viable microorganisms. Microlayers of microbes collect around air bubbles as they rise to the ocean surface, and the microbes are ejected into the air when the bubbles burst (Blanchard and Syzdek, 1972). Algae and protozoa may also be ejected into the atmosphere with seafoam (Schlichting, 1971).

Natural and industrial sources of biogenic aerosols are difficult to separate over the continents. Bovallius et al. (1980) found bacteria in concentrations of 10^4 to 10^5 cm^{-3} over sewage treatment plants. Microorganisms live on our skin and are expelled into the air. The act of changing clothes might propel some 10,000 bacteria per minute into the air. These bacteria are in the size range 0.5 to 2.5 μm radius.

Once in the atmosphere, microorganisms are subject to transport and removal like other aerosol particles. The concentrations of microorganisms depend on location, meteorology, and time of the year. Around cities, the concentrations of bacteria are higher than over the open country and oceans (Bovallius et al., 1980). It has been known for a long time that the annual maximum concentration of bacteria under natural conditions occurs in September and the minimum occurs in March (Buller and Lowe, 1910). A maximum occurs about three hours before midnight, and a minimum in the early morning (Pady and Kramer, 1967). Strong solar radiation reduces the concentration of bacteria (Rüden et al., 1978).

Our knowledge of "dead" biogenic particles in the air is rather crude. Artaxo et al. (1988) state that tropical forests are a major source, but they do not identify the biogenic compounds or quantify their source strengths. Biogenic particles larger than 2 μm radius have been found in higher concentrations in rural than in urban areas (Matthias, 1986). The size distributions of biogenic particles from 0.3 to 50 μm radius is similar to that of the total atmospheric aerosol (see Fig. 2). The biogenic fraction of the total aerosol smaller than 1 μm radius might reach 50% by number; for giant particles it is around 10%.

b.   Volcanoes

While large volcanic eruptions can make major contributions to the stratospheric aerosol reservoir, discharges from most volcanoes remain primarily in the troposphere. Volcanic eruptions usually provide two components for aerosol formation: precursor gases for gas-to-particle conversion (GPC), and water-insoluble dust and ash (silicates and metallic oxides, such as SiO_2, Al_2O_3, and Fe_2O_3). Hofmann (1988) states that most of the primary particles ejected into the atmosphere by volcanoes are too large for long-range transport and settle out quickly, although this does not exclude transport on an intercontinental scale. Thus, Europe has experienced several tropospheric dust veils from Icelandic volcanic eruptions: in 1783 from the volcano Laki (Christ, 1783) and in 1947 from Hekla (Salmi, 1948). If dust from the Sahara travels intercontinentally, why should not volcanic debris? Additionally, released reactive gases that convert rapidly to mostly H_2SO_4 droplets are considered important for the stratosphere, but less so for the troposphere.

Because of their large variations from year to year, the source strength of aerosols from volcanic eruptions is difficult to estimate. Probably the most accurate estimates derive from airborne measurements, such as those described by Hobbs et al. (1982) for Mt. St. Helens. Hobbs et al. (1982) also provide information on particle size distributions and chemical composition of volcanic effluents.

c. Oceans and Fresh Water

The oceans cover approximately 70% of the world's surface. In the past, oceans have been considered the most powerful single source of atmospheric aerosols, up to 10,000 Tg yr^{-1}, mainly in the form of sea salt (Blanchard and Woodcock, 1980). However, this includes the largest particles that are not transported very far. It is not known whether the production of sea-salt aerosols varies by season (see discussion of wind velocity effects below).

The ejection of salt particles into the air, via jet and film drops from air bubble bursting, seems to be well understood. Blanchard and Woodcock (1980) have described four different mechanisms for bringing air into the water: raindrops, snowflakes, supersaturation produced by temperature rises, and whitecaps, with the latter being most effective. Mészáros and Vissy (1974) found cube-shaped particles—classified as sea salt—as small as 0.03 μm, while Junge and Jaenicke (1971) collected particles up to several hundred micrometers. Both authors collected the particles on the Atlantic Ocean under average conditions.

The strength of the oceanic source of salt particles depends on wind speed. Fairall et al. (1983) show a particle-sized resolved vertical volume flux for sea salt as a function of wind speed for ambient relative humidities of 80%. The size range covered is 0.7 to 15 μm radius, and the wind speeds are 6 to 18 m s^{-1}. These results suggest that the particle concentration for the same wind velocity is more than a factor of 10 less over the oceans than over the deserts. This could be due to the energy available at a certain wind velocity. In deserts, particles need only be lifted into the air by the energy of the wind field; over oceans, particles must first be separated from the bulk of the water by bubble bursting and whitecaps and then lofted into the air. An empirical relationship for the mass concentration of salt c (in μg m^{-3}) just above the ocean surface is (Jaenicke, 1988)

$$c = 4.26 \ \exp(0.16u) \tag{1}$$

where u is the wind speed ($1-21$ m s^{-1}). The values for sea salt reported by Fairall et al. (1983) have been compared with data for deserts (see below and Jaenicke, 1988). This analysis reveals that deserts are up to a factor of 100 times more effective as a source of particles (per unit surface area) than oceans. Allowing for uncertainties in the estimates, but keeping in mind the different areal extents, it appears that the oceans and deserts may have about equal source strengths for aerosols in terms of mass per year.

The composition of seawater is remarkably constant, especially the ratio of the major ions to Cl$^-$, but many minor constituents vary with depth (irrelevant for atmospheric aerosol sources) and location (Bowen, 1979). These ions form mainly NaCl, KCl, CaSO$_4$, and (NH$_4$)$_2$SO$_4$. The hygroscopic salts determine the amount of water in individual droplets required for equilibrium with a given relative humidity, and thus the size of the aerosol particles that will be released when the droplets evaporate. Gases released from the oceans, which can serve as precursors for GPC conversion, are discussed later under spatial aerosol sources.

The geographical spread of aerosols from the oceans is difficult to estimate because intervening clouds serve as both sinks and sources of atmospheric aerosols (Dinger et al., 1970; Hegg et al., 1991). (For a summary of aerosol–cloud interaction see Chapter 2.) Hygroscopic particles are found deep into continents (Twomey, 1955). However, most, if not all, of these particles might have been processed by clouds. Neumann (1940) indicated a rapid decrease in concentrations of coastal sea salt with distance inland, unlike aerosols from deserts, which are transported for large distances over the oceans.

Fresh water contains approximately 57.6 mg L^{-1} of dissolved material (Bowen, 1979). The mechanisms for producing aerosol particles from fresh water should be the same as for the oceans. However, the size of the particles released from fresh water should be much smaller, due to the lower amounts of dissolved materials. Aerosols produced from fresh water bodies should be of local influence only, because the surface area of fresh water bodies is much smaller than that of the oceans.

d.   Crustal and Cryospheric Aerosols

On a global scale, deserts are the most important source of particles from the earth's crust. Deserts cover about one-third of the land surface (Pye, 1987). Deserts undergo extreme chemical and physical (temperature as well as mechanical) weathering processes. Thus, unlike the oceans, deserts contain an ample supply of preformed particles. This difference might not have been allowed for in earlier estimates that rated deserts as smaller aerosol sources than the oceans, just because of their smaller area. Not all portions of the deserts are equal aerosol sources. Sand dunes are "dead" surfaces, containing only 4% per mass of particles of the size that can undergo long-range transport in the atmosphere; in dry valleys more than 57% of the particles are in the appropriate size range for long-range transport (d'Almeida and Schütz, 1983).

Particles in soils have often been assumed to be larger than several tenths of a micrometer. Therefore, most measurements in deserts and over barren soil have been tailored for particles of this size (>0.6 μm in Gillette and Walker, 1977). Later studies (d'Almeida and Schütz, 1983) have detected particles as small as 0.01 μm in radius, with the number distribution frequently having a maximum below 0.1 μm radius. Physical weathering alone cannot produce such small particles, but the dissolution of water-soluble minerals might reduce the size of particles (Lerman, 1979; Schroeder, 1985).

The injection of particles into the air from the earth's crust is induced by atmospheric turbulence and triggered by processes like creeping, saltation, and the movement of larger particles. For barren soils, and particles smaller than 10 μm radius, the vertical mass fluxes are 2×10^{-9} and 10^{-6} g cm^{-2} s^{-1} for 6 and 20 m s^{-1} friction velocities, respectively (Gillette, 1981). For crustal-derived aerosols, data covering a wide range of wind velocities is described by

$$c = 52.77 \exp(0.30u) \tag{2}$$

where u is the wind speed (0.5–18 m s^{-1}). In contrast to the oceans, the composition of desert soil varies locally. This offers a potential for source identification by trace elements. However, for the size range of the aerosols involved in long-range transport, there are only minor differences in elemental enrichment factors among all the deserts (Schütz and Sebert, 1987), thus making source identification by this method difficult. Mineral and isotope composition offers a possible hope. Schuetz (1989) found great similarities among most of the soils of the Sahara (except for Northern Saharan soils, which were low in quartz and rich in calcite and palygorskite). Most other minerals are less than 20% of quartz. In the quest for source identifiers, Schuetz mentions illite and kaolinite for Asian dust and material from North America, pyroxene and plagioclase from Mexico, and illite and quartz from the coastal deserts of Peru and Northern Chile (Prospero and Bonatti, 1969). The surface of deserts often contains water-soluble salts.

It now seems likely that the deserts of the world produce about 2000 Tg yr^{-1} of mineral aerosol (d'Almeida, 1986), much more than estimated 20 years ago. It is not known whether the source strength has varied with time. Prospero and Nees (1977) suggested a change in the source strength of the Sahara due to desertification; changes in wind patterns might also cause changes in aerosol production, as well as transport.

Snow surfaces also have preformed particles waiting at the surface for lifting into the atmosphere. The drifting of snow, a process comparable to creeping and saltation in deserts, complicates the estimates of the mass balance of glacial and ice-cap regions, as well as for obtaining unbiased samples of aerosols. Hogan (1975) obtained some preliminary estimates of the latter for particles larger than 100 μm. Since ice crystals derive from water, they contain some hygroscopic substances. Both the north and south polar regions are fenced in rather effectively by meteorological fronts. Polar frontal systems, with their associated precipitation, present a curtain that is probably penetrated by relatively few particles from the polar regions. Thus, the influence of any cryospheric sources should be largely confined to nearby areas.

In this context, the term *resuspension* should be addressed. Pruppacher et al. (1983) give examples that only certain "aerosols" are processed this way. Mostly, these are easy to distinguish, like radioactive aerosols. It seems to be a question of definition of what resuspension really means. The sources "cryosphere" and "biosphere," discussed above, could contribute to resuspension, but only because they are easy to classify. What about deserts? Most material on desert surfaces has been transported (saltation), but only some is catapulted into the atmosphere. What about oceans? Raindrops fall on the oceans and are incorporated into the bulk water. Is the ejection of sea-salt particles into the atmosphere a "resuspension" or a "production?" Thus, in this chapter, resuspension is not considered separately.

e. Biomass Burning

Particles (and gases) from biomass burning have received major attention recently. Soot particles (primary carbon) and fly ash are directly injected into the atmosphere during burning. According to Seiler and Crutzen (1980), 200–450 Tg yr^{-1} are released this way, containing 90–180 Tg yr^{-1} of elemental carbon. It is difficult to define and to separate natural and anthropogenic influences for wildfires, prescribed fires, and agricultural waste burning. Radke et al. (1988) provide emission factors (i.e., mass of material emitted in the atmosphere per unit mass of fuel burned) for particles from a number of biomass fires.

According to Rosen and Novakov (1984), primary carbon consists of two major compounds: graphite (graphitic carbon) and primary organics. Carbon particles are of great importance in the radiation balance of the atmosphere because of the strong absorbing properties of soot. The subject is treated extensively by d'Almeida et al. (1991). Biomass burning also releases large amounts of gases that can form particles in the air by gas-to-particle conversion; these are called secondary organics (secondary carbon). This is a good example of a spatial source, which is the subject of the next section.

2. Spatial Sources

The term *spatial source* refers to particles produced in the atmosphere. Since the precursors are already well distributed, further meteorological mixing processes are not required to achieve widespread spatial distribution of this source of aerosols.

a. Gas-to-Particle Conversion (GPC)

GPC may occur by heterogeneous or homogeneous nucleation. Heterogeneous nucleation refers to the growth of existing nuclei, which serve as a sink for condensable gaseous products. Heterogeneous nucleation can occur with only minor supersaturation of the gases involved. Heterogeneous condensation occurs preferentially on those particles with the largest surface area. For most aerosols (except desert dust—Jaenicke, 1988) such particles are in the radius range $0.1-1$ μm. Homogeneous nucleation forms new particles (smaller than about 0.1 μm). It is also a condensation process. However, to produce new particles, high supersaturations are required. For example, water condenses onto existing cloud condensation nuclei by heterogeneous nucleation at supersaturations not exceeding a few percent. Homogeneous nucleation of water drops, on the other hand, requires more than 300% supersaturation. If two or more condensable species are present (homogeneous heteromolecular nucleation), the nucleation barrier is much lower (e.g., Hegg, 1991).

Three major components are suspected of being involved in GPC: converted sulfates, nitrates, and hydrocarbons. These are discussed in turn below.

i. *Sulfur-Containing Compounds.* Compounds containing gaseous sulfur are released mainly from the biosphere and some 10 to 20% from volcanoes. Such gases are sulfur dioxide (SO_2), hydrogen sulfide (H_2S), carbon disulfide (CS_2), carbonylsulfide (COS), dimethylsulfide (CH_3SCH_3), and dimethyldisulfide (CH_3SSCH_3). The pathways of the non-SO_2 gases in oxidation to SO_2 depend on their photochemical stability. It is rather difficult in the atmosphere to distinguish homogeneous and heterogeneous products. One possibility is to investigate the size-resolved chemical composition of particles.

Winkler (1975) found particles smaller than 0.1 μm radius to consist mainly of sulfuric compounds. In clean air over the oceans, and in remote continental locations, these particles show pronounced daily variations in concentration (Bashurova et al., 1992). This indicates photooxidation production, probably of SO_2 to SO_4^{2-}. However, the photooxidation of SO_2 is a rather slow process. The photooxidation rate increases significantly below the SO_2 dissociation limit at radiation wavelengths near 0.240 μm. Such short wavelengths occur only in the stratosphere. In the troposphere, oxidation may occur with radicals such as OH and HO_2 (Warneck, 1988). The SO_2 pathway also seems to be connected with organic species.

Sulfate particles are concentrated in the size range 0.1 to 1 μm radius (Mehlmann, 1986). Such large particles cannot originate directly from homogeneous GPC. Charlson et al. (1987) suggest that 135 Tg of non-sea-salt SO_4^{2-} are produced annually from dimethylsulfide (DMS) via cloud processes; this would be 50% of the total natural production of sulfate particles (Jaenicke, 1988). So one might conclude that the remainder of 50% of the sulfate particles produced by GPC ends up in the size range below 0.1 μm. However, because of their large mechanical mobility, these particles have an atmospheric residence time of less than 1 day (see below for discussion of residence time) in the troposphere (Jaenicke, 1988), since they coagulate and end up in the 0.1 to 1 μm range.

The transformation of stable sulfur gases (like COS, which has a 44-year lifetime in the troposphere according to Warneck, 1988) to SO_2 and then to SO_4^{2-} in the absence of clouds, together with the direct injection of SO_2 from volcanic outbreaks, is considered to be responsible for the H_2SO_4/H_2O aerosols in the stratosphere (Hofmann, 1988). In the stratosphere, short wavelength radiation can induce the photooxidation of SO_2. Particles produced this way are certainly smaller than 0.1 μm and should play a key role in the formation of polar stratospheric clouds (Hofmann et al., 1990; Peter et al., 1991).

The sulfur source should vary with biogenic activity, which seems to be dominated on a global scale by the northern hemisphere (as the CO_2 annual cycle indicates—Bolin, 1977).

ii. *Nitrogen-Containing Compounds.* In continental aerosols, nitrate particles have radii much larger than 1 μm (Warneck, 1988). Gas-to-particle conversion produces particles only in the size range below 0.1 μm radius. Depending on their concentration, coagulation might eventually transform such small particles into the 0.1 to 1 μm range. Thus, most larger particles can only be the result of evaporated droplets. Consequently, most nitrate-containing particles in the troposphere must result from cloud processes [see (III.A.2.b) below and Chapter 2].

Nitrous oxide (N_2O) is produced principally by microbiological processes in soils and natural water. It is rather stable in the troposphere, but it decomposes chemically in the stratosphere to nitrogen (N_2) and nitric oxide (NO). Nitric oxide is quite rapidly oxidized to nitrogen dioxide (NO_2) by reacting with ozone (O_3). Despite many competing reactions, NO_2 and OH play a role in forming gaseous nitric acid (HNO_3). These reactions have been summarized by Crutzen et al. (1988). Below temperatures of 191 K, the H_2SO_4/H_2O stratospheric aerosols will most likely be solid. $HNO_3 \cdot 3\ H_2O$ molecules condense onto such aerosols to form polar stratospheric clouds. [Some authors prefer to call it *haze,* because these clouds are not composed of water, as earlier anticipated by Steele et al. (1983). However, in cloud physics the term *haze* is used to indicate unactivated cloud droplets—see Chapter 2.] The reduction of $HNO_3 \cdot 3\ H_2O$ molecules opens up the pathway for reactions with chlorine-containing gases to reduce ozone concentrations in the stratosphere. The formation of polar stratospheric clouds is an excellent example of a spatial source and of heterogeneous nucleation. It is now clear that the formation of these clouds depends on temperature and is most effective in the polar regions in winter (see Chapter 8).

iii. *Organic and Carbonaceous Particles.* Natural organic and carbonaceous particles are produced from condensable gaseous precursors released from the biosphere, either as part of their metabolism (e.g., the smell or the blue haze over forest areas—Went, 1964) or in biomass burning (wildfires). However, some contribution certainly comes from volatile compounds of crude oil, leaking to the surface of the earth. Organic materials on a particle not only affect its optical properties, but they can also inhibit the exchange of water vapor between the particle and surrounding gases (Goetz, 1966).

The gaseous and particulate phases of volatile hydrocarbons can exist simultaneously. This coexistence does not necessarily occur at the solution equilibrium for the gaseous and liquid phases. For example, Klippel and Warneck (1980) found 1000 times more formaldehyde on particles than is expected from that equilibrium. On the other hand, Eichmann et al. (1979) extrapolated available vapor pressure data to the ambient temperature range and found the concentration of the higher n-alkenes on aerosol particles to be much lower than anticipated.

Only the higher hydrocarbons are candidates for being present on aerosol particles. $C_{10}-C_{28}$ n-alkanes are readily available in seawater. Hydrocarbons released from the terrestrial biosphere show for n-alkanes a preference of odd over even carbon numbers. Consequently, Warneck (1988) concludes that the oceans are the dominant source of the organic particles and not the terrestrial vegetation, because of the rather even distribution of odd and even numbered n-alkanes in seawater.

Carbonaceous particles seem to be produced primarily by fires—wildfires if we focus on the natural aerosols. Thus, wildfire studies are of great importance for determining gaseous source strengths and the release of soot (Radke et al., 1988). Hoffman and Duce (1977) found most of the mass of organic carbon particles were smaller than 0.25 μm, a clear indication that these particles are formed directly from GPC.

b. Clouds as a Source of Aerosols

Recent studies have indicated that clouds may be a source (as well as a sink) for atmospheric aerosols (e.g., Hobbs, 1986; Hegg et al., 1991; Radke and Hobbs, 1991). This subject is discussed in detail in Chapter 2.

For estimating the strength of clouds as sources for aerosol mass, Pruppacher and Jaenicke (1993) use the following average global assumptions: cloud cover 60%, of which one-half contributes to precipitation ($c_p = 0.3$). The thickness of these clouds is assumed to be $h_p = 3000$ m, with a liquid water content of about $w_p = 0.8 \times 10^{-3}$ kg m^{-3}. It is further assumed that the average global precipitation rate amounts to 1 m per year ($p = 100$ g cm^{-2} yr^{-1}). The residence times of liquid water in clouds τ_p is then found to be

$$\tau_p = \frac{c_p h_p w_p}{p} = 2.3 \times 10^4 \text{ s} = 6.3 \text{ h} \tag{3}$$

This implies that the considered precipitating cloud layer "empties" about every 6 hours. Since on global average one may assume that those clouds have a residence time of about $\tau_{cc} = 1$ h, one may conclude that the cloud has to form six times; it evaporates five times and precipitates the sixth time. The precipitation over the whole globe is 5×10^{17} kg yr^{-1}. Six times this amount evaporates in the air, leaving behind its chemical mass in the form of aerosol. From the literature, the chemical mass in cloud water is 1.25 mg L^{-1} (e.g., Hegg and Hobbs, 1982). Hence, evaporating clouds could "produce" ~3000 Tg yr^{-1} of atmospheric aerosols. This source could be responsible for the rather uniform background of aerosols in the troposphere, which will be discussed later.

c.   Extraterrestrial Sources

Cosmic dust contributes to the stratospheric aerosols and less to the tropo-sphere. Other extraterrestrial particles might be produced from abrasions of me-teorite showers that enter the upper atmosphere. Extraterrestrial particles received considerable attention in the earlier literature (e.g., Thomsen, 1953), but con-temporary source strength estimates have reduced it from the early estimates of 2 Tg yr^{-1} of magnetic spherules to a negligible 1.4–1.8 Gg yr^{-1} (d'Almeida et al., 1991).

3.   Summary of Source Strengths

Table 1 lists our best estimates of the source strengths of aerosols from natural origins. However, many of these estimates derive from the 1970s and have not been updated. Clouds are a newly recognized aerosol source; indeed, they may be the major natural source of aerosols larger than 0.1 μm radius.

B.   Particle Size Distributions and Chemical Compositions

In recent years many measurements have been obtained of the aerosol size distri-bution under various conditions. Occasionally the instruments used render inter-pretation difficult. For example, measurements obtained from optical counters with a monochromatic (laser) light source usually show a number size distribution with a depression in the large particle size range. This is due to the scattering function of monochromatic light (Kim and Boatman, 1990) and its associated am-

Table 1

Estimates of the Global Strengths of Aerosol Particles from Natural Sources[a]

Source	Strength (Tg yr^{-1})
(a) Widespread surface sources	
Oceans and fresh water bodies	~1000–2000
Crust and cryosphere	~2000–
Biosphere and biomass burning	~450–
Volcanoes	~15–90
(b) Spatial sources	
Gas-to-particle conversion	~1300
Clouds	~3000
Extraterrestrial	~10

[a]The data have been taken mainly from Jaenicke (1978b, 1988), Bach (1976), Hidy and Brock (1971), Schütz (1980), and d'Almeida et al. (1991). The estimates are rather crude; this is indicated by the open upper limit and by the approximation sign "~".

biguity in estimating the particle size. This ambiguity is assumed to cancel out (Leaitch and Isaac, 1991). However, recent numerical simulations show that such counters seriously distort depression-free aerosol size distributions and tend to produce artificial modes (Hanusch and Jaenicke, 1993). Such instruments, however, might provide reliable measurements of the total particle concentrations in certain particle size ranges. White-light optical particle counters can also produce erroneous results, but not mode-like structures in the size distribution.

For many purposes [e.g., cloud models (Flossmann, 1991; Chuang, 1991), deposition calculations (Kramm, 1991), and radiation models (d'Almeida et al., 1991)], up to three lognormal size distributions have been used to describe atmospheric aerosol size distributions:

$$\frac{dN(r)}{d(\log r)} = \sum_{i=1}^{3} \frac{n_i}{\sqrt{2\pi} \log \sigma_i} \exp\left\{ -\frac{(\log r/R_i)^2}{2(\log \sigma_i)^2} \right\} \tag{4}$$

where r is the particle radius (in μm), $N(r)$ is the cumulative particle number distribution (in cm^{-3}) for particles larger than r, R_i is the mean particle radius (in μm), n_i is the integral of the ith normal function, and $\log \sigma_i$ is a measure of particle polydispersity. In this case, up to nine parameters are needed to describe the aerosol size distribution. Table 2 lists values for these parameters for several types of tropospheric aerosols (Jaenicke, 1988).

In Table 2 the tropospheric aerosols are divided into the following types.

Polar aerosols. These refer to aerosols close to the surface in the Arctic and Antarctica. Only a few measurements are available (Jaenicke and Schütz, 1982; Ito, 1982; Shaw, 1986). The distribution reflects the aged character of these aerosols. The parameter values given in Table 2 are supported by recent evaluations of many years of observation (Jaenicke et al., 1992).

Background aerosols. These refer to the tropospheric aerosols above cloud layers. Most measurements have been carried out at mountain stations or in subsiding airmasses that reflect mid-tropospheric conditions (e.g., Mészáros and Vissy, 1974; Tymen et al., 1975; Prahm et al., 1976; Jaenicke, 1978a, 1979; Schultz et al., 1978; Reiter et al., 1984). Measurements in the free troposphere (e.g., from airplanes) are limited.

Maritime aerosols. Maritime aerosols are expected to be composed of the background aerosol plus sea salt. The parameters given in Table 2 are for ambient relative humidities and a wind force of 4 (5.5–7.9 m s^{-1}) (Chaen, 1973; Mészáros and Vissy, 1974; d'Almeida and Schütz, 1983). Because of the fitting procedure, the relation "maritime aerosol − background aerosol = sea salt aerosol" is not always fulfilled. Recent measurements over the tropical and South Pacific oceans confirm the fine structure of the aerosol size distribution in the radius range below 0.1 μm (Hoppel and Frick, 1990).

Remote continental aerosols. These should reflect aerosols close to the sur-

Table 2
Parameters for Models of Aerosol Size Distributions Described by
the Sum of Three Lognormal Functions

Aerosol	Range[a]	i	n_i (cm^{-3})	R_i (μm)	log σ_i
Polar	I	1	2.17×10^1	0.0689	0.245
	II	2	1.86×10^{-1}	0.375	0.300
	III	3	3.04×10^{-4}	4.29	0.291
Background	I	1	1.29×10^2	0.0036	0.645
	II	2	5.97×10^1	0.127	0.253
	II	3	6.35×10^1	0.259	0.425
Maritime	I	1	1.33×10^2	0.0039	0.657
	II	2	6.66×10^1	0.133	0.210
	II	3	3.06×10^0	0.29	0.396
Remote continental	I	1	3.20×10^3	0.01	0.161
	I	2	2.90×10^3	0.058	0.217
	II	3	3.00×10^{-1}	0.9	0.380
Desert dust storm	I	1	7.26×10^2	0.001	0.247
	I	2	1.14×10^3	0.0188	0.770
	III	3	1.78×10^{-1}	10.8	0.438
Rural	I	1	6.65×10^3	0.00739	0.225
	I	2	1.47×10^2	0.0269	0.557
	I	3	1.99×10^3	0.0419	0.266
Urban	I	1	9.93×10^4	0.00651	0.245
	I	2	1.11×10^3	0.00714	0.666
	I	3	3.64×10^4	0.0248	0.337

Source: Mainly from Jaenicke, 1988.
[a] I, R_i is the Aitken or nucleation particle mode ($0.001 < R_i < 0.1$ μm); II, R_i is in the large or accumulation particles mode ($0.1 < R_i < 1$ μm); III, R_i is in the giant or coarse particles mode ($R_i > 1$ μm). For other explanations see Eq. (4).

face in continental regions not greatly affected by humans. Such measurements are rather sparse. Recent measurements in Siberia (Koutsemogii, 1992) report the parameters given in Table 2.

Desert dust storm aerosols. These natural aerosols might be considered an extreme. Because of the strength and extent of its source and its potential for long-range transport, it influences vast areas and deserves separate treatment (Schütz and Jaenicke, 1974; Jaenicke and Schütz, 1978; d'Almeida and Schütz, 1983). The total number concentration in such aerosols is rather low (~ 1500 cm^{-3}), but its mass load is rather high.

Rural aerosols. The aerosols in rural areas are mainly continental but with moderate impact from anthropogenic sources (Jaenicke and Junge, 1967; Sverdrup, 1977; Tanner and Marlow, 1977; Whitby, 1978; Reiter et al., 1984). The measurements of Hobbs et al. (1985) in the lower troposphere over the high plains of North America reflect the model distributions for rural and urban aerosols given in Table 2, rather than the background or remote continental aerosol model distributions.

Urban aerosols. Because of the dominant influence of anthropogenic emissions from industries, home heating, and traffic, the aerosols in cities and human settlements have a special character. Surprisingly, only a few measurements exist (Jaenicke and Junge, 1967; Whitby, 1978).

Shown in Fig. 1 are size distributions of six aerosol types. The radius range is limited to particles larger than 0.001 μm. However, for many of the distributions, the measurements close to 0.001 μm are uncertain due to instrumentation problems. The main differences among the curves are for the smallest and the largest particles. This probably reflects the influence of particle residence time (see below). The only exception is the desert dust storm. However, this aerosol is closest to its source. Atmospheric aerosols show only limited concentration variations in the range 0.1 to 1 μm. Particles in this size range influence atmospheric visibility and cloud formation. Visibility (without precipitation) varies from about 1 km to several hundred kilometers, or more than two and a half orders of magnitude. That variation is reflected exactly by variations in the appropriate size range of the aerosol model size distributions shown in Fig. 1. The models show a pronounced multimodal structure, particularly in the mass or volume distributions (Fig. 2).

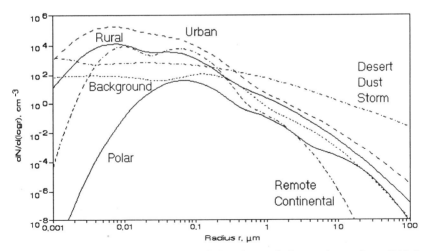

Figure 1   Model number size distributions of selected atmospheric aerosols according to Table 2.

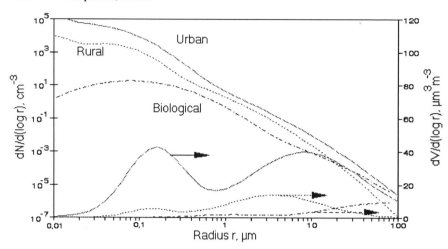

Figure 2   Model description of the number and volume distributions of biological aerosols. The upper curves are number distributions in logarithmic scale (left side); the corresponding lower curves are volume distributions in linear scale (right side). The rural and urban aerosols are the aerosol model distributions of Table 2.

However, as Table 2 indicates, the mean radii of these "modes" usually do not coincide with the size ranges often used for classification.

Aerosols produced from the biosphere need special attention. While the concentrations of pollen and bacteria have been investigated extensively in the field of hygiene, measurements of the size distributions of all biological particles in the tropospheric aerosols are virtually absent. Figure 2 shows some recent preliminary averaged results of measurements in the size range 0.2 to 80 μm radius for selected seasons in a rural to urban environment in Central Europe (Matthias-Maser, 1992). This study seems to support an external mixture of the aerosols, probably because it was collected close to its sources. However, for other particles, the removal of biogenic particles from the atmosphere certainly has an influence on their distributions downwind of sources.

Although abundant data are available on the elemental compositions of atmospheric aerosols (Artaxo et al., 1988), there is less information on ionic compositions, for example, for urban aerosols. Table 3 summarizes some of the available data. Urban aerosols are given for three locations: Tees-side, an industrial area on the NE coast of England (Eggleton, 1969); West Covina, in greater Los Angeles, California (Stelson and Seinfeld, 1981); and Sapporo in Japan (Ohta and Okita, 1990). Remote continental aerosols are expected to be present at Wank Mountain (1780 MSL) in the Bavarian Alps (Reiter et al., 1984). Polar aerosols are represented by two data sets from the Arctic (Heintzenberg et al., 1981; Li and

Winchester, 1989). The Atlantic Ocean data are compiled from various sources (Warneck, 1988): Na^+, K^+, Ca^{2+}, Mg^{2+}, SO_4^{2-}, Cl^- (Buat-Ménard et al., 1974); Na^+, K^+, Ca^{2+}, Mg^{2+}, Fe_2O_3 (Hoffman et al., 1974); Na^+, SO_4^{2-}, NH_4^+, NO_3^- (Gravenhorst, 1975 a, b); and Br^- (Pacific Ocean, Duce et al., 1965).

The total organic mass (TOM) component of atmospheric aerosols is determined only occasionally. It is included in Table 3 by using the ratio of EEOM/TOM (EEOM is ether-extractable organic matter) of 0.5 from Eichmann et al. (1980). The TOM for polar aerosols (Lannefors et al., 1983) assumes the measured carbon content to be half of the TOM.

Matthias-Maser (1992) determined the volume fraction of biological aerosols to be 14.6% of the total aerosols. If we follow Gregory (1973), the bulk density of biological material is close to 1 g cm^{-3}, which is comparable to that generally assumed for the tropospheric aerosols. So the volume fraction of the biological aerosols must be close to its mass fraction. Using these numbers the masses of biological material have been estimated in Table 3. It seems as if one-fifth of the mass fraction of aerosol has been neglected in the past.

If aerosol volume distributions are weighted by particle bulk density (in the presence of hygroscopic salts and atmospheric water a particle density of 1 g cm^{-3} seems reasonable) and then integrated, a mass concentration results. This does not always compare well with the values given in Table 3. This disagreement reflects uncertainties and variances in current estimates.

C.   Transport, Geographical Distribution, Residence Time, and
     Influence of Clouds

There is overwhelming evidence that tropospheric aerosols can be transported over large distances, even intercontinentally (e.g., Hobbs et al., 1971). One of the best and longest documented cases is that of Saharan dust (summarized in Schutz et al., 1990). Thus, desert aerosols can cover a much larger area than the deserts themselves (see Fig. 3). For example, Saharan dust extends far into the Atlantic Ocean and into Asia. Even the rain forest of South America may be nourished by minerals in transported and precipitated dust from the Saharan desert (Swap et al., 1992). Arctic haze is the result of long-range transport from continents (Stonehouse, 1986).

The vertical distribution of aerosols in the troposphere needs additional investigation. Even though some data (e.g., Hobbs et al., 1985) are available, the measurements of particle size distribution are sparse. For most cases only the vertical distribution of selected parameters is available. Typically these parameters show an exponential decrease (or increase) with altitude z up to a certain height (Gras, 1991) and a rather constant profile above. Such a distribution can be described, in a generalized way, by the following relation:

Table 3

Mass Concentrations (in $\mu g\ m^{-3}$) of the Tropospheric Aerosols by Chemical Composition

Aerosol:	Urban			Remote continental	Polar		Maritime
Location:	Tees-side England	West Covina California	Sapporo Japan	Wank Mt. Germany	Ny-Alesund Spitsbergen	Barrow Alaska	Atlantic Ocean
Year:	1967	1973	1982	1972–1982	1979	1986	pre 1975
SO_4^{2-}	13.80	16.47	2.8–5.3	2.15	2.32	1.91	2.58
NO_3^-	3.00	9.70	0.1–1.6	0.85	0.055	0.13	0.050
Cl^-	3.18	0.73	0.1–1.6	0.087	0.013	1.11	4.63
Br^-	0.07	0.53				0.05	0.015
NH_4^+	4.84	6.93	0.6–1.8	1.00	0.23	0.65	0.16
Na^+	1.18	3.10	0.3–0.9	0.047	0.042	0.68	2.91
K^+	0.44	0.90		0.045	0.023	0.38	0.11
Ca^{2+}	1.56	1.93	0.2–1.0	0.082	0.073		0.17
Mg^{2+}	0.60	1.37			0.032		0.40

18

Al₂O₃	3.63	6.43		0.20			
SiO₂	5.91	21.10		0.51			
Fe₂O₃	5.32	3.83		0.10	0.24		0.14
CaO	—[a]	—[a]		0.09	0.91		
PbO				0.020			
ZnO				0.020			
Cu				0.002			
Cd				0.001			
Totals[b]	43.53	75.70	32.3	48.73	3.94	4.87[c]	11.17
TOM[d]	56		2.4–16.0	4	0.14		2
Biological material[e]	17.0	22.5	7.3				

Source: References are given in the text.

[a] All Ca assumed to be water soluble.

[b] Totals are the individual compounds added up.

[c] The total is from Li and Winchester (1989) who also include Methanesulfonate (0.012), Formate⁻ (0.24), Acetate⁻ (0.73), Propionate⁻ (0.147), and Pyruvate⁻ (0.010); adding the column gives a greater value.

[d] TOM is total organic matter, calculated from carbon content.

[e] Biological material is estimated for rural and urban aerosols only, using the fraction 14.6% of the total mass (measurements of Matthias-Maser, 1992).

19

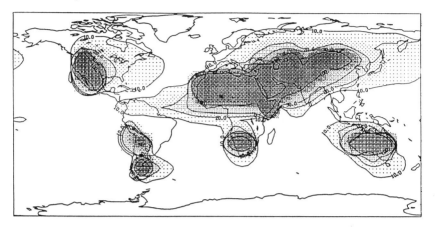

Figure 3   Mass concentration of mineral aerosols (in μg m⁻³) for April at the 1000 hPa pressure level (Wefers and Jaenicke, 1990). This distribution was obtained (Wefers, 1990) by adapting a three-dimensional climatological transport model (MOGUNTIA, Zimmermann et al., 1989) for the desert aerosols. The aerosols in that model are followed in 10 size ranges between 0.1 and 100 μm radius. The activities of the sources are calculated from wind velocities. The sinks include only the settling velocity of the aerosols. Thus the model is more realistic in dry than in wet geographical regions.

$$p = p_0 \left\{ \exp\left(\frac{-z}{|H_p|}\right) + \left(\frac{p_B}{p_0}\right)^v \right\}^v ; \quad H_p \neq 0; \quad v = \frac{H_p}{|H_p|} \quad (5)$$

where p_0 is the concentration at the surface, p_B is the concentration of the background aerosol aloft, and H_p is the scale height. By replacing p with n and m in Eq. (5), one obtains the vertical distributions of the aerosol number and mass concentrations, respectively. Parameters are listed in Table 4.

For mass concentration, usually $p_B = 0$. This converts Eq. (5) to the more familiar exponential decrease typical for atmospheric pressure:

$$m = m_0 \exp\left(-\frac{z}{H_m}\right) \quad (6)$$

Figure 4 shows vertical profiles of aerosol mass concentration compiled from many publications.

For maritime aerosols, Kristament et al. (1992) recently published two different particle size ranges that reflect the range of the particle mass concentration. Their data have been summarized explicitly with a scale height $H_m = 900$ m (the $m_0 = 16$ μg m⁻³ were taken from Table 2) up to 2400 m and a constant mass above.

A similar behavior for continental aerosols is reflected in Gillette and Blifford's (1971) observations and Junge's (1963) conversion to "large" particle concentration using radiation measurements from the 1950s. An $H_m = 730$ m ($m_0 = 20$ μg m⁻³ from Table 2) up to 2400 m has been estimated from Junge (1963) for

Table 4

Parameters of Vertical Aerosol Profiles[a]

Aerosol type	Altitude (m)	Scale height H_p (m)	Surface value p_0	Background value p_B	Main references
		Aerosol mass concentration ($p = m$, in μg m^{-3})			
Ocean	→2400	900	16	0	Kristament et al. (1992)
Remote continental	→2400	730	20	0	Jung (1952a), Gillette and Blifford (1971)
Desert	→6000	2000	150	0	Model only: Wefers (1990)
Polar	→6000	30000	3	0	Leaitch and Isaac (1991)
Background	→tropopause	∞	1	0	Leaitch and Isaac (1991)
		Aerosol number concentration ($p = n$, in cm^{-3})[b]			
Ocean	→1000	−290 to 440	10–3000	300	Gras (1991), Bigg et al. (1984)
Remote continental	→2500	1100 to 550	3000–30000	300	Jung (1963), Rosen et al. (1978)
Polar	→500	−130	10	200	Rosen et al. (1978), Heintzenberg et al. (1991)
	500→ tropopause	∞	200	200	
Background	→tropopause	∞	300–2000	300	Rosen et al. (1978), Gras (1991)

[a] The parameters are explained in Eq. (5). Replace the general term p by n or m to get mass or number concentration, respectively:

$$p = p_0 \left\{ \exp\left(\frac{-z}{|H_p|}\right) + \left(\frac{p_B}{p_0}\right)^v \right\}^{1/v}; \quad H_p \neq 0; \quad v = \frac{H_p}{|H_p|}.$$

[b] The order of the H_n values corresponds to the order of the n_0 values. The minus sign is intended.

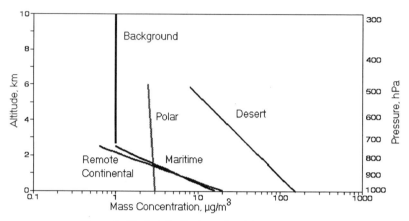

Figure 4   Vertical distributions of particle mass concentration.

remote continental aerosols.

Climatological, three-dimensional, global model calculations by Wefers (1990) for desert aerosols in April in the northern hemisphere must be used as a substitute for observations: $H_m = 2000$ m with $m_0 = 150$ μg m^{-3} up to 6000 m. This reflects high-reaching atmospheric convection in these regions.

The vertical profile for polar aerosols shown in Fig. 4 has been taken from Leaitch and Isaac (1991). They calculated vertical profiles of mixing mass ratios up to 6000 m from particle size distribution measurements assuming a particle density of 2 g cm^{-3}. With a more realistic particle density of 1 g cm^{-3} their results can be represented by $H_m = 30$ km and a surface mass concentration of 3.0 μg m^{-3}.

Data on background aerosols have been summarized by Leaitch and Isaac (1991); they are included in Fig. 4 as 1 μg m^{-3} in the appropriate altitude range. While the maritime and remote continental aerosols fit nicely with the background aerosol above about 2 km, the polar and desert aerosols do not seem to approach the background value above. (However, a discrepancy of a factor of 2 in measurements could result in an agreement.)

Arctic haze needs special attention. As Shaw (1983) indicates, this aerosol is enhanced above ambient values in mass but not in number. These hazes generally form in laminated, elevated layers in late winter and spring (e.g., Rahn et al., 1977; Brock et al., 1989). The graphitic carbon concentration in these layers at altitudes above 1 km can be close to surface values in cities like Denver, Colorado (Rosen and Hansen, 1986). Measurements of optical thickness (Leiterer and Graeser, 1990) point to even greater carbon concentrations at times in the Arctic than in Central Europe. The nature and concentrations of the particles comprising arctic haze have been discussed by Radke et al. (1989) and Ishizaka et al. (1989). It is believed that the long-range transport responsible for arctic haze is associated with anticyclonic conditions during winter, with transport of air pollutants from industrialized centers in northern Europe, North America, and Asia to the Arctic (Stonehouse, 1986; Brock et al., 1989).

Vertical distributions of aerosols number concentrations can be represented by Eq. (5) by replacing p with n. In Fig. 5 the background aerosols again seem to be rather well-defined and almost omnipresent. In interpreting these results it should be kept in mind that aerosol number concentrations have a large variability. This is due in large part to the short atmospheric residence times of small particles that dominate number concentrations (see number size distribution in Fig. 1 and the discussion of residence time below). The background aerosols seem to exhibit an almost constant vertical profile, with concentrations in the range 300 to 3000 cm^{-3} above the mixing layer up to the tropopause. However, individual profiles may show increases or decreases in concentration with height (Rosen et al., 1978; Gras, 1991).

The concentrations of polar aerosols are more stable, with values around 200 cm^{-3} up to the tropopause, but often with much lower concentrations at the surface (Rosen et al., 1978; Heintzenberg et al., 1991).

Remote continental aerosols generally show a decrease in number concentration with altitude in the mixed layer. The steepness of this decline depends on the surface concentrations (Junge, 1963; Rosen et al., 1978). Above the mixed layer, continental aerosol concentrations approach the background aerosol con-

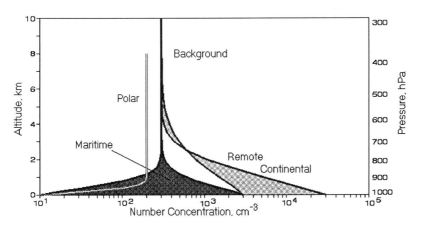

Figure 5 Vertical distribution of aerosol number concentration. The ranges of the number concentrations are shown for ocean as well as remote continental aerosols. For the sake of a graphically smoothed presentation, the following equation has been used (where n_0 is the number concentration at the surface and n_B is the number concentration of the corresponding background aerosols aloft):

$$n = n_0 \left\{ \exp\left(\frac{-z}{|H_n|}\right) + \left(\frac{n_B}{n_0}\right)^v \right\}^v ; \qquad H_n \neq 0; \quad v = \frac{H_n}{|H_n|}$$

centration.

Over the oceans there is an almost constant aerosol profile concentration (Bigg et al., 1984; Hegg et al., 1991; Gras, 1991), which confirms the idea that the background aerosols reach to the ocean surface. In the mixed layer, the maritime aerosols at times behave similarly to the polar aerosols, in that a lower concentration is observed at the surface than for the background aerosols aloft. In this case a negative scale height H_n results. On other occasions, the concentration of aerosols is higher near the surface, and consequently H_n is positive.

Although the vertical profiles of aerosol number concentrations can be variable, some recent observations indicate systematic deviations, especially in the vicinity of clouds. Figure 6 shows three typical profiles. For low ambient aerosol concentrations over the Pacific Ocean, Hegg et al. (1991), Hegg (1991), and Radke and Hobbs (1991) observed local maxima in concentrations near clouds (see also Pueschel and Livingston, 1990). They explained this by homogeneous, heteromolecular nucleation of sulfuric acid at high relative humidities. In a low ambient aerosol concentration, Albrecht et al. (1988) attributed an increase in concentrations above a marine stratus cloud ("brown layers") to the influence of distant continental sources. Over Europe, Weber (1990) reported a large reduction in aerosol concentrations between continental cumulus clouds as compared to the concentrations above and below clouds.

Much of the variations in tropospheric aerosol concentrations is the result of their residence time. The residence time τ depends on the compositions of the

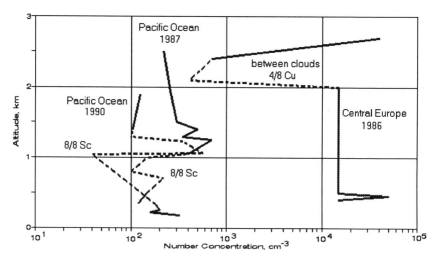

Figure 6   Individual vertical profiles of Aitken and condensation nucleus concentrations. The dashed profiles indicate the extent of the cloud layers. Pacific Ocean data of 1990 are adapted from Hegg (1991); Pacific Ocean data of 1987 are adapted from Albrecht et al. (1988); Central Europe data of 1986 are taken from Weber (1990).

aerosols, the particle radius r, and geographic location. In a generalized way it can be represented by (Jaenicke, 1988)

$$\frac{1}{\tau} = \frac{1}{K}\left(\frac{r}{R}\right)^2 + \frac{1}{K}\left(\frac{r}{R}\right)^{-2} + \frac{1}{\tau_{wet}} \tag{7}$$

where K is a constant (1.28×10^8 s); R is the radius of the particles with the maximum residence time ($R = 0.3$ μm); and τ_{wet} is a measure of the influence of wet removal ($\tau_{wet} = 8$ days in the lower troposphere and $\tau_{wet} = 3$ weeks in the middle to upper troposphere). This expression yields a maximum residence time for particles with a radius of 0.3 μm of about one week in cloudy areas and up to several weeks in the upper troposphere. Interestingly, the particle range affected by clouds exhibits the longest residence time in the atmosphere. This maximum residence time is limited through wet removal processes. Again, this indicates the inefficiency of wet removal processes. Smaller and larger particles are suspended for much shorter times in the air. The short residence time of smaller particles is the result of Brownian motion and coagulation, which transforms the particles into larger sizes. In a chemical sense, this does not limit their residence time, because the chemical substance is not removed from the atmosphere. However, if we look at how aerosols act in the atmosphere, we see the importance of their size range. The short residence time of the larger particles results from their larger settling velocities.

Roughly speaking, aerosol residence times are comparable to those of water and sulfuric acid. Such short residence times greatly increase the spatial and temporal variability of aerosols, especially in the vicinity of nonuniform sources. Longer aerosol residence times in the upper troposphere, together with the spatial source "clouds," appear to be the reasons for the comparatively constant background of aerosols.

IV. Some Outstanding Problems

A few of the outstanding problems in tropospheric aerosol studies are discussed briefly below.

- The extreme variability of tropospheric aerosols gives rise to many problems and questions. Because of the short residence time of tropospheric aerosols (the residence time of water is comparable to that of aerosols), measuring networks with the density of precipitation networks are necessary to understand atmospheric aerosols. Sound strategies are needed to obtain the most useful and reliable measurements, particularly as dense measuring networks are unlikely to become available. Because of this variability it was recommended in a recent meeting (in Garmisch-Partenkirchen, 1992) that the World Meteorological Organization's BAPMon stations reduce the upper particle size included in mass concentration measurements, rather than increase the number of stations.

- There are problems associated with instrumentation. Aerosols encountered in industrial processes and hygiene are quite often different from atmospheric aerosols. Atmospheric aerosols are usually more complex chemically and are present in much lower concentrations. Consequently, when used for atmospheric measurements, instruments developed for industrial use often have to be used at their lowest operational range. Many of the instruments and techniques that have been optimized for atmospheric aerosol research are not generally available, particularly in remote areas where measurements are sorely needed. Also, many research instruments are one of a kind. There is a need for more intercomparisons and exchanges of instruments. A start has been made in a series of workshops for the intercomparison of Aitken and ice nuclei counters organized by the Committee on Nucleation and Atmospheric Aerosols and also by the CODATA workshop (Jaenicke and Hahn, 1989).

- The atmospheric aerosol community needs to devise improved methods for presenting results and findings for a highly complex system in a multidimensional array. For example, do sulfate aerosols, in the strict sense of the term, exist, or are all aerosols mixed in nature? What is the equivalent sulfate particle size for a mixed aerosol and how is it best represented graphically? What are the size ranges of areosols on which traces of sulfate are present?

- There is evidence that tropospheric aerosols are changing in many respects. However, there is no evidence that aerosols on a global scale are increasing in mass due to human activities. For example, if we include the new estimates of the aerosol source strength of clouds, the anthropogenic contribution to atmospheric aerosols is reduced in proportion. There are hints that atmospheric chemistry may be changing, particularly for minor constituents with a rather long residence time. Consequently, there is a need to establish and obtain long time-series measurements. However, this should not be solely for the purposes of monitoring, but mainly for basic understanding. Meteorologists have long realized the need to obtain certain measurements over long time periods in order to establish climatic variations. Atmospheric aerosols need similar attention and treatment.

- The establishment of the International Global Aerosol Program—IGAP (Deepak and Vali, 1991) may help solve some of the problems listed above.

Acknowledgment

Professor P. V. Hobbs has spent a large effort in helping to design this chapter.

References

Aitken, J. (1888). *Nature* **37**, 428–430.

Albrecht, B. A., Randall, D. A., and Nicolls, S. (1988). *Bull. Am. Meteor. Soc.* **69**, 618–626.

Artaxo, P., Storms, H., Bruynseels, F., van Grierke, P., and Maenhaut, W. (1988). *J. Geophys. Res.* **93**, 1605–1615.

Bach, W. (1976). *Geophys. Space Phys.* **14**, 429.

Bashurova, V. S., Dreiling, V., Hodger, T. V., Jaenicke, R., Koutsenogii, K. P., Koutsenogii, P. K., Krämer, M., Makarov, V. I., Obolkin, V. A., Potjomkin, V. L., and Pusep, A. Y. (1992). *J. Aerosol Sci.* **23**, 191–199.

Bigg, E. K., Gras, J. L., and Evans, C. (1984). *J. Atmos. Chem,* **1**, 203–214.

Blanchard, D. C. and Syzdek, L. D. (1972). *J. Geophys. Res.* **77**, 5087–5099.

Blanchard, D. C., and Woodcock, A. H. (1980). *Ann. N. Y. Acad. Sci.* **338**, 330–347.

Bolin, B. (1977). *Science* **196**, 613–615.

Bovallius, A., Roffey, R., and Hemmingson, E. (1980). *Ann. N. Y. Acad. Sci.* **353**, 186–200.

Bowen, H. J. M. (1979). "Environmental Chemistry of the Elements." Academic Press, London.

Brock, C. A., Radke, L. F., Lyons, J. H., and Hobbs, P. V. (1989). *J. Atmos. Chem.* **9**, 129–148.

Buat-Ménard, P., Morelli, J., and Chesselet, R. (1974). *J. Rech. Atmos.* **8**, 661.

Buller, A. H. R., and Lowe, C. W. (1910). *Transactions Roy. Soc. Canada* **4**, 41–58.

Chaen, M. (1973). *Memoirs Faculty Fisheries Kagoshima University* **22**, 49–107.

Chagnon, C. W., and Junge, C. E. (1965). "The Size Distribution of Radioactive Aerosols in the Upper Troposphere." *Environ. Res. Papers* **139**.

Chamberlain, A. C., and Dyson, E. D. (1956). *Brit. J. Radiol.* **29**, 317–325.

Charlson, R. J., Lovelock, J. E., Andreae, M. O., and Warren, S. G. (1987). *Nature* **326**, 655–661.

Christ, J. E. (1783). "Von der außerordentlichen Witterung des Jahrs 1783, in Ansehung des anhaltenden und heftigen Hoehenrauchs; Vom Thermometer und Barometer, von dem natürlichen Barometer unserer Gegend, dem Feldberg oder der Hoehe, und von der Beschaffenheit und Entstehung unserer gewoehnlichen Lufterscheinungen, wie auch etwas von den Erdbeben." Hermannische Buchhandlung, Frankfurt and Leipzig.

Chuang, C. C. (1991). *5th Intern. Conf. Precip. Scav. Atmos.-Surf.- Exchange Proc., Richland*, 127.

Cour, P., and van Campo, M. (1980). *C. R. Acad. Sci. Paris* **290**, 1043–1046.

Crutzen, P. J., Brühl, C., Schmailzl, U., and Arnold, F. (1988). *In* "Aerosols and Climate" (P. V. Hobbs and M. P. McCormick, eds.), 287–304. Deepak Publishing.

d'Almeida, G. A. (1986). *J. Clim. Appl. Meteor.* **25**, 903–916.

d'Almeida, G. A., Koepke, P., and Shettle, E. P. (1991). "Atmospheric Aerosols. Global Climatology and Radiative Characteristics." Deepak Publishing.

d'Almeida, G. A., and Schütz, L. (1983). *J. Clim. Appl. Meteor.* **22**, 233–243.

Deepak, A., Vali, G. (1991). "The International Global Aerosol Program (IGAP) Plan: Overview." ICCP/IRC/ICACGP.

Dinger, J., Howell, H., and Wojciechowski, T. (1970). *J. Atmos. Sci.* **27**, 791–797.

Dingle, A. N. (1966). *J. Rech. Atmos.* **2**, 231–237.

Duce, R. A., Winchester, J. W., and Van Nahe, T. W. (1965). *J. Geophys. Res.* **70**, 1775–1799.

Eggleton, A. E. J. (1969). *Atmos. Environ.* **3**, 355–372.

Eichmann, R., Ketseridis, G., Schebeske, G., Jaenicke, R., Hahn, J., Warneck, P., Junge, C. (1980). *Atmos. Environ.* **14**, 695–703.

Eichmann, R., Neuling, P., Ketseridis, G., Hahn, J., Jaenicke, R., and Junge, C. (1979). *Atmos. Environ.* **13**, 587–599.

Fairall, C. W., Davidson, K. L., and Schacher, G. E. (1983). *Tellus* **35B**, 31–39.

Flossmann, A. I. (1991). *Tellus* **43B**, 301–321.

Flossmann, A. I., and Pruppacher, H. R. (1988). *J. Atmos. Sci.* **45**, 1857–1871.

Flossmann, A. I., Pruppacher, H. R., and Topalian, J. H. (1987). *J. Atmos. Sci.* **44**, 2912–2923.

Gillette, D. (1981). *In* "Desert Dust: Origin, Characteristics, and Effect on Man" (T. L. Pewe, ed.), Geol. Soc. of America Special Paper 186.

Gillette, D. A., and Blifford, I. H. (1971). *J. Atmos. Sci.* **28**, 1199–1210.

Gillette, D. A., and Walker, T. R. (1977). *Soil Science* **123**, 97–110.

Goetz, A. (1966). "Microphysical and Chemical Aspects of Sea Fog and Oceanic Haze," Final Report US Navy Weather Research Facility Norfolk Contract No. N189(188)59510A.

Gras, J. L. (1991). *J. Geophys. Res.* **96**, 5345–5356.

Gravenhorst, G. (1975*a*). "Der Sulphatanteil im atmosphärischen Aerosol über dem Nordatlantik." Berichte des Instituts für Meteorologie und Geophysik der Universität Frankfurt 30.

Gravenhorst, G. (1975*b*). *Meteor Forsch.-Erg.* **B10**, 22–33.

Gregory, P. H. (1973). "The Microbiology of the Atmosphere," Leonard Hill Books, Aylesbury, 377 pp.

Grosjean, D., Lewis, R., Fung, K., Swanson, R., and Countess, R. (1982). *75th Annual Meeting of the Air Pollution Control Association, New Orleans.*

Hanusch, T., and Jaenicke, R. (1993). *Aerosol Science and Technology* (in press).

Hegg, D. A. (1991). *Geophys. Res. Lett.* **18**, 995–998.

Hegg, D. A., and Hobbs, P. V. (1982). *Atmos. Environ.* **16**, 2663–2668.

Hegg, D. A., Radke, L. F., and Hobbs, P. V. (1991). *J. Geophys. Res.* **96**, 18727–18733.

Heintzenberg, J., Hansson, H. C., and Lannefors, H. (1981). *Tellus* **33**, 162–171.

Heintzenberg, J., Ström, J., and Ogren, J. A. (1991). *Atmos. Environ.* **25A**, 621–627.

Hidy, G. M., and Brock, J. R. (1971). *Proc. 2nd Int. Clean Air Congress*, 1088 pp.

Hirano, S. S., Maher, E. A., Kelman, A., and Upper C. D. (1978). *In* "Station de Pathologi végétate et Phytobacteriologi," 717–724.

Hobbs, P. V. (1986). *In* "Chemistry of Multiphase Atmospheric Systems" (W. Jaeschke, ed.), 191–211, NATO ASI Series G6, Springer-Verlag.

Hobbs, P. V., Bluhm, G., and Ohtake, T. (1971). *Tellus* **23**, 28–39.

Hobbs, P. V., Bowdle, D. A., and Radke, L. F. (1985). *J. Clim. Appl. Meteor.* **24**, 1344–1356.

Hobbs, P. V., and Radke, L. F. (1992). *Science* **256**, 987–991.

Hobbs, P. V., Tuell, J. P., Hegg, D. A., Radke, L. F., Eltgroth, M. W. (1982). *J. Geophys. Res.* **87**, 11062–11086.

Hoffman, E. J., and Duce, R. A. (1977). *Geophys. Res. Lett.* **4**, 449–452.

Hoffman, E. J., Hoffman, G. L., and Duce, R. A. (1974). *J. Rech. Atmos.* **8**, 676.

Hofmann, D. J. (1988). *In* "Aerosols and Climate" (P. V. Hobbs and M. P. McCormick, eds.), 195–214, Deepak Publishing.

Hofmann, D. J., Deshler, T., Arnold, F., and Schlager, H. (1990). *Geophys. Res. Lett.* **17**, 1279–1282.

Hogan, A. W. (1975). *J. App. Meteor.* **14**, 428–429.

Hoppel, W. A., and Frick, G. M. (1990). *Atmos. Environ.* **24A**, 645–659.

Ishizaka, Y., Hobbs, P. V., and Radke, L. F. (1989). *J. Atmos. Chem.* **9**, 149–159.

Ito, T. (1982). *Antarctic Record* **76**, 1–19.

Jaenicke, R. (1978*a*). *Meteor Forsch. Erg.* **B13**, 1–9.

Jaenicke, R. (1978*b*). *Pageoph* **116**, 283–292.

Jaenicke, R. (1978*c*). *Atmos. Environ.* **12**, 161–169.

Jaenicke, R. (1979). *J. Aerosol Sci.* **10**, 205–207.

Jaenicke, R. (1988). *In* "Numerical Data and Functional Relationships in Science and Technology," Landolt-Börnstein New Series, V: Geophysics and Space Research, 4: Meteorology (G. Fischer, ed.) *b:* Physical and Chemical Properties of the Air. 391–457, Springer, Berlin.

Jaenicke, R., Dreiling, V., Lehmann, E., Koutsenoguii, P. K., and Stingl, J. (1992). *Tellus* **44B**, 311–317.

Jaenicke, R., and Hahn, J. (1989). *In* "Directions for Internationally Compatible Environment Data. A CODATA Workshop." International Council for Scientific Unions (IUGG), Committee on Data for Science and Technology, Volume II. (G. C. Carter, ed.), 1–111.

Jaenicke, R., and Junge, C. (1967). *Beitr. Phys. Atmosph.* **40**, 129–142.

Jaenicke, R., and Schütz, L. (1978). *J. Geophys. Res.* **83**, 3585–3599.
Jaenicke, R., and Schütz, L. (1982). *J. Hungarian Meteorol. Service* **86**, 235–241.
Junge, C. E. (1952*a*). *Ann. Meteorologie Beiheft,* 1–55.
Junge, C. (1952*b*). *Berichte des Deutschen Wetterdienstes US-Zone* **35**, 261–277.
Junge, C. E. (1961). *J. Meteor.* **18**, 501–509.
Junge, C. (1963). *J. Rech. Atmos.* **1**, 185–188.
Junge, C., and Jaenicke, R. (1971). *J. Aerosol Sci.* **2**, 305–314.
Kim, Y. J., and Boatman, J. F. (1990). *J. Atmos. Oceanic Techn.* **7**, 681–688.
Klippel, W., and Warneck, P. (1980). *Atmos. Environ.* **14**, 809–818.
Kramm, G. (1991). *Meteor. Rundsch.* **43**, 65–80.
Kristament, I. S., Liley, J. B., and Harvey, M. J. (1992 submitted). *J. Geophys. Res.*
Koutsemogii, P. (1992). "Measurements of Remote Continental Aerosol in Siberia." Ph. D. Dissertation University Mainz.
Lannefors, H., Heintzenberg, J., and Hansson, H. C. (1983). *Tellus* **35B**, 40–54.
Leaitch, W. R., and Isaac, G. A. (1991). *Atmos. Environ.* **25A**, 601–619.
Leiterer, U., and Graeser, J. (1990). "Arctic Haze" -März 1989. *Z. Meteorol.* **40**, 64–66.
Lerman, A. (1979). *In* "Geochemical Processes; Water and Sediment Environments." John Wiley & Sons.
Levin, Z., and Yankofsky, S. A. (1988). *In* "Atmospheric Aerosol and Nucleation" (P. E. Wagner and G. Vali, eds.), Springer-Verlag, Berlin, Heidelberg.
Levin, Z., Sadlerman, N., Moshe, A., Bertold, T., and Yankofsky, S. A. (1980). *Proc. 8th Int. Conf. on Cloud Physics, Clermont Ferrand, France.*
Li, S. M., and Winchester, J. W. (1989). *Atmos. Environ.* **23**, 2387–2399.
Lindemann, J., Constantinidou, H., Barchet, W. R., and Upper, C. D. (1982). *App. Environ. Microbiology* **44**, 1059–1063.
Maki, L. R., and Willoughby, K. J. (1978). *J. App. Meteor.* **17**, 1049–1053.
Matthias, S. (1986). "Die biogene Komponente im atmosphärischen Aerosol im Mainzer Raum." Master Thesis Institut für Meteorologie, University Mainz.
Matthias-Maser, S. (1992). "Entwicklung einer Methode zur Identifizierung von biologischen Aerosolpartikeln mit Radius $r > 0.2$ μm zur Bestimmung ihrer atmosphärischen Größenverteilung." Ph.D. Dissertation University Mainz.
Mehlmann, A. (1986). "Größenverteilung des Aerosolnitrates und seine Beziehung zur gasförmigen Salpetersäure." Ph.D. Dissertation University Mainz.
Mészáros, A., and Vissy, K. (1974). *J. Aerosol Sci.* **5**, 101–109.
Mie, G. (1908). *Ann. Phys.* **25**, 377–445.
Monahan, E. C., and Muircheartaigh, I. O. (1980). *J. Phys. Ocean.* **10**, 2094–2099.
Neumann, H. R. (1940). *Gerlands Beiträge Geophysik* **56**, 49–91.
Nolan, P. J., and Doherty, D. J. (1950). *Proc. Roy. Irish Acad.* **A53**, 163–179.
Ohta, S., and Okita, T. (1990). *Atmos. Environ.* **24A**, 815–822.
Orsini, C. Q., Tabacniks, M. H., Artaxo, P., Andrade, M. F., and Kerr, A. S. (1986). *Atmos. Environ.* **20**, 2259–2269.
Pady, S. M., and Kramer, C. L. (1967). *Mycologica* **59**, 714–716.
Peter, T., Brühl, C., and Crutzen, P. J. (1991). *Geophys. Res. Lett.* **18**, 1465–1468.
Prahm, L. P., Torp, U., and Stern, R. M. (1976). *Tellus* **28**, 355–372.
Prospero, J. M., and Bonatti, E. (1969). *J. Geophys. Res.* **74**, 3362–3371.
Prospero, J. M., and Nees, R. T. (1977). *Science* **196**, 1196–1198.
Pruppacher, H. R., and Jaenicke, R. (1993). Submitted to *Tellus.*
Pruppacher, H. R., Semonin, R. G., and Slinn, W. G. N. (1983). "Precipitation Scavenging, Dry Deposition and Resuspension—Volume 2: Dry Deposition and Resuspension," Elsevier.
Pueschel, R. F., and Livingston, J. M. (1990). *In* "Proceedings of the Third Chinese Aerosol Conference, Beijing, China, September 29–October 1, 1990," The Institute of Aerosol Science and Technology, Chinese Society of Particuoligy, 182–184.

Pye, K. (1987). "Aeolian Dust and Dust Deposits." Academic Press, New York.

Radke, L. F., Brock, C. A., Lyons, J. H., and Hobbs, P. V. (1989). *Atmos. Envir.* **23**, 2417–2430.

Radke, L. D., Hegg, D. A., Lyons, J. H., Brock, C. A., Hobbs, P. V., Weiss, R., and Rasmussen, R. (1988). *In* "Aerosols and Climate" (P. V. Hobbs and M. P. McCormick, eds.), 411–422, Deepak Publishing.

Radke, L. F., and Hobbs, P. V. (1991). *J. Atmos. Sci.* **48**, 1190–1193.

Rahn, K. A., Borys, R. D., Shaw, G. E. (1977). *Nature* **268**, 713–715.

Reiter, R., Carnuth, W., and Sladkovic, R. (1984). *Arch. Met Geoph. Biokl.* **B35**, 179–201.

Rosen, H., and Hansen, A. D. A. (1986). *In:* "Arctic Air Pollution." (B. Stonehouse, ed.) 101–120, Cambridge University Press.

Rosen, H., and Novakov, T. (1984). *In* "Aerosols and Their Climatic Effects" (H. E. Gerber and A. Deepak, eds.), 83–94, Deepak Publishing, Hampton.

Rosen, J. M., Hofmann, D. J., and Kaselau, K. H. (1978). *J. Appl. Meteor.* **17**, 1737–1740.

Rüden H., Thofern, E., Fischer, P., and Mihm, U. (1978). *Pageoph* **16**, 335–350.

Salmi, M. (1948). *C. R. Soc. Geol. Finland* **21**.

Savoie, D. L., and Prospero, J. M. (1982). *Geophys. Res. Lett.* **9**, 1207–1210.

Schlichting, H. E. (1971). *Botanica Marina* **14**, 24–28.

Schnell, R. C. (1982). *Tellus* **34**, 92–95.

Schnell, R. C., and Vali, G. (1976). *J. Atmos. Sci.* **33**, 1554–1564.

Schroeder, J. H. (1985). *J. African Earth Sci.* **3**, 371–380.

Schuetz, L. (1989). *In* "Paleoclimatology and Paleometeorology: Modern and Past Patterns of Global Atmospheric Transport" (M. Leinen and M. Sarntheim, eds.), 359–383, Kluwer Academic Publishers.

Schultz, A., Fritz, G., and Schumann, G. (1978). *Meteor Forsch.-Erg.* **B13**, 10–13.

Schütz, L. (1980). *Ann. N.Y. Acad. Sci.* **338**, 515–532.

Schütz, L., and Jaenicke, R. (1974). *J. Appl. Meteorol.* **13**, 863–870.

Schütz, L., and Sebert, M. (1987). *J. Aerosol Sci.* **18**, 1–10.

Schutz, L. W., Prospero, J. M., Buat-Ménard, P., Carvalho, R. A. C., Cruzado, A., Harriss, R., Heidam, N. Z., and Jaenicke, R. (1990). *In* "The Long-Range Atmospheric Transport of Natural and Contaminant Substances." NATO ASI Series C: Mathematical and Physical Sciences 297 (A. H. Knap, ed.), 197–230, Kluwer Acad. Publ.

Seiler, W., and Crutzen, P. J. (1980). *Climatic Change* **2**, 207–247.

Shaw, G. E. (1983). *J. Atmos. Sci.* **40**, 1313–1320.

Shaw, G. E. (1986). *J. Aerosol Sci.* **17**, 937–945.

Steele, H. M., Hamill, P., McCormick, M. P., and Swisler, T. J. (1983). *J. Atmos. Sci.* **40**, 2055–2067.

Stelson, A. W., and Seinfeld, J. H. (1981). *Environ. Sci. Technol.* **15**, 671–679.

Stonehouse, B. (1986). "Arctic Air Pollution." Cambridge University Press.

Sverdrup, G. M. (1977). "Parametric Measurements of Submicron Atmospheric Aerosol Size Distributions," Progress Report EPA Research Grant R 803851-02, Particle Technology Laboratory Publication Number 320.

Swap, R., Garstang, M., and Greco, S. (1992). *Tellus* **44B**, 133–149.

Tanner, R. I., and Marlow, W. H. (1977). *Atmos. Environ.* **11**, 1143–1150.

Thomsen, W. (1953). *Sky and Telescope,* 147–148.

Twomey, S. (1955). *J. Meteor.* **12**, 81–86.

Twomey, S. (1959). *Geofis. Pura Appl.* **43**, 227–242.

Tymen, G., Butor, J. F., Renoux, A., and Madelaine, G. (1975). *Chemosphere* **4**, 357–360.

Vali, G., Christensen, M., Fresh, R. W., Galyvan, E. L., Maki, R. L., and Schnell, R. C. (1976). *J. Atmos. Sci.* **33**, 1565–1570.

Wagman, J., Lee, R. E., and Axt, C. J. (1967). *Atmos. Environ.* **1**, 479–489.

Warneck, P. (1988). "Chemistry of the Natural Atmosphere." Academic Press.

Weber, S. (1990). "Messung von vertikalen Profilen der Aerosolteilchen-Gesamtkonzentration mit einem Motorsegler im Rhein-Nahe-Gebiet bei ausgesuchten konvektiven Wetterlagen." Master Thesis, University Mainz.

Weber, S., and Jaenicke, R. (1988). *In* "Atmospheric Aerosols and Nucleation." Lecture Notes in Physics 309 (P. E. Wagner and G. Vali, eds.), 56–60, Springer-Verlag, Berlin.

Wefers, M. (1990). "Numerische Simulation der globalen 3-dimensionalen Verteilung arider Aerosolpartikel ohne Berücksichtigung der nassen Deposition." Master Thesis, University Mainz.

Wefers, M., and Jaenicke, R. (1990). *In* "Aerosols. Proc. Third International Aerosol Conference Kyoto Japan" (S. Masuda and K. Takahashi, eds.), 1086–1089, Pergamon Press.

Went, F. W. (1964). *Proc. National Acad. Sci.* **51,** 1259–1267.

Whitby, K. T. (1973). *In* "VIII International Conference on Nucleation Leningrad."

Whitby, K. T. (1978). *Atmos. Environ.* **12,** 135–159.

Winkler, P. (1975). *Geophys. Res. Lett.* **2,** 45–48.

Woodcock, A. H., and Gifford, M. M. (1949). *J. Marine Res.* **8,** 177–197.

Yankofski, S. A., Levin, Z., Bertold, T., and Sandlerman, N. (1981). *J. Appl. Meteor.* **20,** 1013–1019.

Zimmermann, P., Feichter, J., Rath, H. K., Crutzen, P. J., and Weiss, W. (1989). *Atmos. Envir.* **23,** 25–35.

Chapter 2 | Aerosol–Cloud Interactions

Peter V. Hobbs
Atmospheric Sciences Department
University of Washington
Seattle, Washington

Aerosol–cloud–climate interactions involve the six interactions depicted schematically in the preface to this book. This chapter is concerned with a review of two of these interactions: the effects of atmospheric aerosol on clouds, and the effects of clouds on atmospheric aerosol. The discussion is confined to tropospheric clouds; aerosol–cloud interactions in the stratosphere are discussed in Chapter 8.

Various aspects of the effects of aerosol on clouds have been studied for many years. This is because the aerosol on which cloud droplets form determine the initial concentrations and sizes of the droplets. Aerosol may also play a role in the formation of ice in clouds. Thus, through their effects on both the nature (water or ice) and the size distribution of cloud particles, aerosol can play a role in determining whether or not clouds precipitate and the radiative properties of clouds. These topics are reviewed in the first part of this chapter.

Clouds and precipitation are important sinks for atmospheric aerosol. This affects both the size distribution and chemical nature of atmospheric aerosol, as well as the chemical composition of clouds and precipitation. In addition to modifying existing aerosol, some recent research indicates that clouds can be involved in the nucleation of new aerosol. These topics form the subject of the second part of this chapter.

I. Effects of Aerosol on Clouds

A. Cloud Condensation Nuclei, Cloud Droplet Spectra, and Precipitation

1. Heterogeneous Nucleation of Water Vapor Condensation

The classical laboratory experiments of Coulier (1875) and Aitken (1923) demonstrated that, in particle-free air, water vapor condenses (by *homogeneous* nucleation) to form droplets only if the supersaturation is several hundred per-

Aerosol–Cloud–Climate Interactions   33

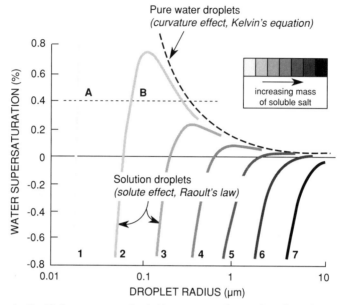

Figure 1   Equilibrium supersaturation (with respect to a plane surface of water) over droplets of pure water (dashed curve), and over droplets containing fixed masses of dissolved salt, as a function of droplet radius.

cent.[1] By contrast, in the presence of aerosol, such as those that exist in the atmosphere, water droplets form (by *heterogeneous* nucleation) at supersaturations of about 2% (102% relative humidity) or less.

The theory of heterogeneous nucleation onto particles is well known (see, for example, Mason, 1971) and need not be repeated here. Instead, we will emphasize the results of the theory, which are important for understanding the effects of aerosol on cloud droplet formation. The essential facts are summarized in Fig. 1, in the form of the Köhler (1921) curves, which shows the equilibrium supersaturation (and therefore equilibrium vapor pressure) over droplets of various sizes and containing various masses of dissolved salt. The Köhler curves incorporate the following effects:

- As the size of a droplet increases, the equilibrium supersaturation above its surface decreases (Kelvin's equation). In the case of a pure water droplet, the effect is shown by the dashed curve in Fig. 1: the smaller the droplet, the larger the equilibrium vapor pressure. The curves for droplets containing fixed masses of salt approach the Kelvin curve as they increase in size, since the droplets become increasingly dilute solutions.

[1] Supersaturation (in %) $= (e/e_s - 1)100$, where e is the vapor pressure of the air and e_s is the saturated vapor pressure over a plane surface of water.

- The equilibrium supersaturation is lowered by dissolved solutes (Raoult's law). For droplets containing fixed masses of solutes, this effect becomes increasingly strong the smaller the droplet. Thus, as can be seen from Fig. 1, for very small solution droplets the equilibrium vapor pressure is less than that over a plane surface of water (i.e., the equilibrium supersaturation is negative).

In addition to giving the equilibrium supersaturation just above the surface of a droplet, the Köhler curves can be used to determine how droplets will grow in an environment at a fixed supersaturation. For example, in an environment with a supersaturation of 0.4%, water-insoluble particles with a radius less than about 0.5 μm could not serve as nuclei for the growth of droplets because the equilibrium supersaturation of a droplet that forms on them is initially >0.4% (the dashed curve in Fig. 1). However, a water-insoluble particle with a radius >0.5 μm could serve as the nucleus for water-vapor condensation at a supersaturation of 0.4%.

In this same environment, with a supersaturation of 0.4%, droplets containing dissolved solute (of dry radius much less than 0.5 μm) represented by curve 1 in Fig. 1, would increase in size up to point A, at which point the droplet would be in equilibrium with the environmental vapor pressure. Similarly, droplets represented by curve 2 would grow to point B. In the atmosphere, droplets that are in equilibrium states such as A and B (to the left of the peaks in their Köhler curve) are referred to as *unactivated drops* or *haze*. Haze can significantly decrease the intensity of solar radiation reaching the earth; it can also cause sharp decreases in visibility. Note that water-soluble particles can form haze at vapor pressures below water saturation.

Next consider a solution droplet represented by curve 3 in Fig. 1, which is exposed to an environmental supersaturation of 0.4%. In this case, because the peak in the Köhler curve lies below 0.4% supersaturation, the droplet can grow by condensation up the left-hand side of the Köhler curve, over the peak in this curve, and down the right-hand side of the curve. The droplet is now said to be *activated*, because it has formed a cloud drop many micrometers in radius. In general, a water-soluble particle will be activated in an environment with supersaturation S if $S > S_c$, where S_c is the peak value of the supersaturation given by the Köhler curve for the particle. S_c depends on the number of soluble ions in the particle (e.g., Hudson and Clarke, 1992):

$$S_c = \frac{2.5 \times 10^5}{(\text{number of soluble ions})^{0.5}} \tag{1}$$

It should be noted that the Köhler curves represent equilibrium conditions. Large particles have large equilibrium radii and may have insufficient times to grow to their equilibrium sizes in clouds with strong updrafts. Consequently, they may remain unactivated (e.g., Jensen and Charlson, 1984). However, since the radii of such particles are large, they are already essentially cloud drops; therefore, the fact that, strictly speaking, they are not activated has no practical significance.

2. Cloud Condensation Nuclei

It can be seen from the above discussions and Fig. 1 that, provided they are large enough, water-insoluble particles and smaller water soluble particles can serve as *cloud condensation nuclei* (CCN) at supersaturations of a few tenths of one percent; the larger the particle, and the more soluble ions it contains, the lower the critical supersaturation at which it can serve as a CCN. Such particles provide CCN for cloud formation in the earth's atmosphere.

Cloud condensation nucleus spectra (i.e., concentrations of CCN as a function of supersaturation) can be measured by exposing samples of air to known supersaturations and then measuring the concentrations of droplets that form (see, for example, Wallace and Hobbs, 1977). Such measurements show that the concentration of CCN (in cm^{-3}) active at a supersaturation S (in %) can often be fitted to an expression of the form (Twomey, 1959):

$$n = cS^k \qquad (2)$$

It should be noted that measurements of the parameters c and k in Eq. (2) bypass the need for any chemical information on CCN, at least as far as cloud microphysics is concerned. This is because Eq. (2) implicitly contains all the chemical and particle size information. Twomey (1959) gives average values of $c = 310$ and $k = \frac{1}{3}$ for marine air and $c = 600$ and $k = \frac{2}{5}$ for continental air. Hegg and Hobbs (1992) have reviewed more recent measurements of CCN in marine air. They conclude that the median values of c and k are 200 and $\frac{1}{2}$, respectively, and that about half of the particles in clean marine air are active as CCN at a supersaturation of 1%. Twomey and Wojciechowski (1969) give $c = 600$ and $k = \frac{1}{2}$ for continental air.

3. Development of Cloud Droplet Spectra

Calculations of the growth of cloud droplets by condensation in a parcel of air rising adiabatically and with a constant (specified) speed is straightforward (Howell, 1949). Some results of such calculations for marine air are shown in Fig. 2; these illustrate the following important points.

- As a parcel of air is lifted it is cooled adiabatically, and this increases its relative humidity (RH). After the parcel reaches its lifting condensation level (RH = 100%, $S = 0$%), the supersaturation initially increases linearly (dashed line in Fig. 2).
- As the supersaturation increases, various particles in the air parcel serve as CCN (e.g., lines 3–5 in Fig. 2).
- As droplets grow by condensation they increasingly offset the increase in supersaturation due to cooling. Consequently, the supersaturation reaches a

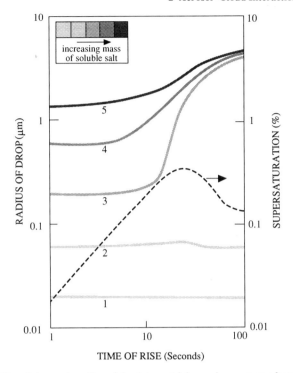

Figure 2 Growth by condensation of droplets containing various masses of water-soluble salts in a parcel of air moving upward with a speed of 60 cm s^{-1} (lines 1–5, referenced to left-hand ordinate). Also shown is the supersaturation of the rising air parcel (dashed line, referenced to right-hand ordinate). The calculations are for marine air. (Adapted from Howell, 1949.)

peak value and then declines (dashed line in Fig. 2). This peak value, which depends on the spectrum of CCN and the updraft speed of the air (Twomey, 1959), is typically <2%. For the same updraft speed, the peak supersaturation is greater in marine clouds (which have fewer CCN) than in continental clouds.

- Even though the CCN that are activated have different sizes, the droplets that form on them quickly approach a rather uniform radius (see right-hand side of curves 3–5 in Fig. 2). This is because the radius of the smaller droplets increases faster than the radius of the larger droplets ($dr/dt \propto 1/r$; see, for example, Wallace and Hobbs, 1977). Thus, this simple model predicts the development of a monodispersed cloud droplet size spectrum.

- For the case study depicted in Fig. 2, the peak supersaturation was not sufficient to activate some of the smaller soluble particles (e.g., curves 1 and 2). Initially, condensation occurred on these particles to form haze, but when the supersaturation decreased the water on these haze particles evaporated.

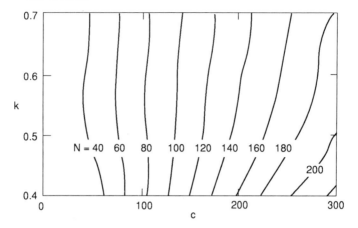

Figure 3   Relationship between the parameters c and k in Eq. (2) and cloud droplet concentrations (N in cm^{-3}) for a range of vertical air speeds typical of marine stratus clouds. (From Hegg et al., 1991a.)

While approximate analytical relationships suggest that cloud droplet concentrations depend on both c and k in Eq. (2) (e.g., Squires, 1952; Twomey, 1959), more sophisticated numerical cloud models (Hegg et al., 1991a) show that the k dependence can be quite weak (Fig. 3)—see also Bigg (1990).

It is important to emphasize that well above cloud base, processes other than those considered in the above simple model can play an important role in determining cloud droplet size spectra. For example, the mixing of dry ambient air into a cloud (a nonadiabatic process) can modify the droplet size spectra (e.g., Bower and Choularton, 1988). If a few droplets grow large enough they can start to collide with and collect smaller droplets (e.g., Pruppacher and Klett, 1978). Also, if ice particles begin to form in a cloud they will grow at the expense of cloud droplets. The simple model described above applies best to regions near cloud base and to relatively thin layer clouds. Shown in Fig. 4 are some recent measurements in shallow marine stratus clouds, which demonstrate that the mean droplet concentration increases as the concentration of CCN active at 1% supersaturation [i.e., the value of c in Eq. (2)] increases, as predicted by the simple adiabatic model.

The number of CCN that are activated (and therefore the number of cloud droplets) increases as the peak supersaturation in the air increases (Fig. 1). Even though, for the same updraft speed, the peak supersaturations in marine clouds are greater than those in continental clouds, the simple model described above predicts that continental clouds will contain larger droplet concentrations than marine clouds because of the higher concentrations of CCN in continental air. Therefore, since the liquid water contents of continental and marine clouds do not differ greatly, the average size of the droplets in continental clouds should be less than

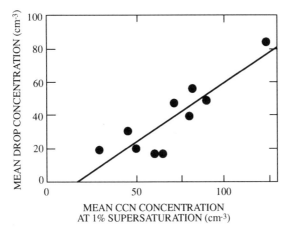

Figure 4 Measurements of mean droplet concentrations versus concentrations of CCN active at 1% supersaturation in marine stratus clouds. (Adapted from Hegg et al., 1991*a*.)

in marine clouds. These predictions from the simple adiabatic model are supported by observations. For example, marine cumulus clouds have a median droplet concentration of ~45 cm^{-3}, a rather broad droplet size spectrum, and a median droplet diameter of about 30 μm; continental cumuli have a median droplet concentration of ~230 cm^{-3}, narrow droplet size spectra, and a median diameter near 10 μm (Squires, 1956, 1958*a,b*; Pruppacher and Klett, 1978; Hobbs and Rangno, 1990*a,b*, 1992).

The effects of anthropogenic CCN can be seen in the microstructures of clouds. For example, because forest fires are sources of CCN they increase droplet concentrations, and decrease the width of droplet size spectra, in clouds downwind (Eagan et al., 1974). Paper mills are prolific sources of large (approximately 0.1–1 μm diameter) and giant (>1.0 μm diameter) CCN, but not of small (<0.1 μm diameter) CCN (Hobbs et al., 1970; Hindman et al., 1977*a,b*). Thus, the number concentrations of cloud droplets with diameters ≥30 μm are raised downwind of paper mills (Hindman et al., 1977*b*).

4.   Effects of Aerosol on Development of Precipitation

For a cloud to rain, about one cloud droplet in a million has to grow to precipitable size so that its radius increases from about 10 μm to 1000 μm. Growth of cloud droplets by vapor deposition can produce droplets with radii of 10 μm or so, but thereafter growth by this mechanism is slow (Fig. 2). For warm clouds (i.e., clouds containing no ice), drops somewhat larger than average can grow to precipitation size only by collecting smaller droplets that lie in their fallpath. Clearly, a broad droplet size distribution is more conducive to such growth than a narrow distribution. As we have seen, due to the lower concentrations of CCN in marine air, marine clouds tend to have broader droplet size distributions than their continental counterparts. Thus, other things being equal (e.g., for similar updraft speeds and cloud depths), marine clouds produce precipitation more effectively than continental clouds.

As we will see in Section I.C, aerosol also affect the development of ice in clouds and, therefore, the production of precipitation by ice processes.

5. Stability of CCN Populations

Baker and Charlson (1990) suggested that the boundary layer over the ocean, which is often topped by a stratiform cloud layer, is bistable with respect to CCN populations. In their model, when the source strength of CCN was low enough, drizzle maintained low CCN concentrations (~ 10 cm^{-3}). They identified this stable state with very clean marine air, in which the production of CCN was <0.002 cm^{-3} s^{-1}. When the production of CCN, and therefore cloud droplets, were slightly higher, their calculations showed that drizzle ceased and the CCN concentrations stabilized at ~ 1000 cm^{-3} (a high value compared to observations). They suggested that under these conditions the main sink of CCN is dry particle coagulation.

Using a more sophisticated model, Ackerman et al. (1992) found that the marine atmosphere does not exhibit such an instability. They found that the instability postulated by Baker and Charlson (1990) resulted from their oversimplified treatment of cloud microphysical processes, particularly the production of drizzle. As shown in Fig. 5, the model of Ackerman et al. produces a smooth transition between the low and high particle concentration states predicted by Baker and Charlson. This is in contrast to the step-function response of the model of Baker and Charlson. The unstable positive feedback loop between droplet concentration and drizzle rate predicted by Baker and Charlson is not reproduced by Ackerman et al. for at least two reasons. First, at high droplet-number concentrations the peak supersaturations are reduced, and therefore not all droplets are activated (a negative feedback loop). Second, at low droplet-number concentrations, Baker and Charlson assign an unrealistically high collection efficiency to describe gravitational collisions between droplets.

The application of these ideas to marine stratiform clouds may be limited by the long time scales needed for equilibrium to be reached, at least for the case of high CCN concentrations. If marine clouds require five or more days to establish equilibrium under high CCN production rates, it is unlikely that such a balance will ever be reached. This is because atmospheric processes that operate on shorter time scales, such as advection of CCN or frontal passages, will tend to disrupt the evolution toward a steady state. Finally, it should be noted that these model simulations do not take into account that clouds are not only sinks but also sources of CCN (see Sections II.B and II.D). Inclusion of this additional source, which is time dependent, may prevent the system from ever reaching a steady state.

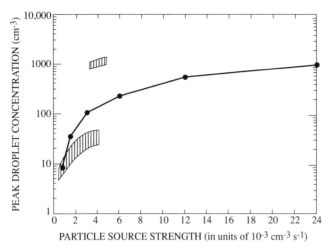

PARTICLE SOURCE STRENGTH (in units of 10^{-3} cm^{-3} s^{-1})

Figure 5   A comparison of the results from the steady-state model of Baker and Charlson (1990) (hatched regions) with results from the model of Ackerman et al. (1992) (solid line).

These recent studies emphasize the need for more research on the sinks of CCN, as well as the sources.

B.   Aerosol Effects on Cloud Radiative Properties [2]

Photographs of the earth from satellites show that the brightest objects, and therefore the major reflectors of the sun's radiation, are clouds and snow-covered surfaces. Therefore, if atmospheric aerosol modify the radiative properties of clouds, they have the potential for large effects on the earth's radiation balance. In this section we first review some basic concepts relevant to this issue. We then discuss the potential for increasing atmospheric aerosol concentrations to affect the radiative properties of clouds and therefore global climate. Several examples of such modifications by anthropogenic aerosol are described. The section concludes with a discussion of a hypothesis for a planetary thermostat involving aerosol–cloud interactions.

1.   Effects of Aerosol on Cloud Optical Thickness and Albedo

The optical thickness (τ) of a cloud of depth h containing number concentration $n(r)$ of droplets of radius r is given by (e.g., Twomey, 1977)

$$\tau = \pi h \int_0^\infty Q_e r^2 n(r) \ dr \tag{3}$$

[2] It should be emphasized that we are concerned here with the effects of aerosol on the radiative properties of clouds (i.e., the so-called *indirect* radiative effects of aerosol). There is mounting evidence that aerosol may also have important *direct* effects on the radiation balance of the earth; this topic is the subject of Chapter 3.

where Q_e is the extinction efficiency. At visible wavelengths, and when $\lambda \ll r$, Q_e has a value close to 2. Therefore, if we assume a narrow droplet-size spectrum with a mean radius $\bar{r}$, Eq. (3) simplifies to

$$\tau = 2\pi h(\bar{r})^2 N \tag{4}$$

where N is the total number concentration of droplets:

$$N = \int_0^\infty n(r) \, dr \tag{5}$$

The total liquid water content (W) of the cloud is

$$W = \frac{4}{3}\pi\rho_L \int_0^\infty r^3 n(r) \, dr \tag{6}$$

where ρ_L is the density of liquid water, or

$$W = \frac{4}{3}\pi\rho_L(\bar{r})^3 N \tag{7}$$

From Eqs. (4) and (7),

$$\tau = 2.4\left(\frac{W}{\rho_L}\right)^{2/3} hN^{1/3} \tag{8}$$

It follows from Eq. (8) that if the liquid water content and depth of a cloud are constant,

$$\frac{\Delta\tau}{\tau} = \frac{1}{3}\frac{\Delta N}{N} \tag{9}$$

The *albedo* (or *reflectance*) of a cloud is the fraction of the incident radiation that is reflected by the cloud integrated over the hemisphere of backscattering. The large areas of rather thin stratiform clouds that cover an appreciable fraction of the earth's surface have an albedo of about 0.50. Therefore, the energy balance of the earth is rather sensitive to the albedo of these clouds.

The albedo A of a cloud is given to a good approximation by (Meador and Weaver, 1980)

$$A = \frac{(1 - g)\tau}{1 + (1 - g)\tau} \tag{10}$$

where g is the scattering asymmetry factor, which is the average value of the cosine of the scattering angle. For the scattering of solar radiation by clouds, $g \simeq 0.85$. Hence, Eq. (10) becomes

$$A \simeq \frac{\tau}{\tau + 6.7} \tag{11}$$

Provided the liquid water content and cloud depth are held constant, Eqs. (9) and (11) yield

$$\frac{\Delta A}{\Delta N} = \frac{A(1 - A)}{3N} \tag{12}$$

It follows from Eq. (12) that for a given N, $\Delta A/\Delta N$ has a maximum value when $A = 0.5$, although the curve is rather flat for a range of values on either side of $A = 0.5$. For a fixed value of A, $\Delta A/\Delta N$ is inversely proportional to N. Hence, as can be seen from Fig. 6, the albedo A of a cloud is most sensitive to changes in N when A has values from about 0.25 to 0.75 and N is small (i.e., in air with low CCN concentrations). The maximum value of $\Delta A/\Delta N$ approaches 1% per additional cloud drop per cubic centimeter of air! As Twomey (1991) points out, to produce a change in CCN of 1 cm^{-3} from the surface of the earth up to a height of 1 km over the whole globe would require only about 50 tonnes of material. Assuming a CCN residence time of just 2 days, this perturbation in CCN concentration would require an injection rate of 1–10 kilotonnes of material per year. Since this is small compared to anthropogenic sulfur emissions ($\sim$6 megatonnes per year), anthropogenic emissions should be affecting cloud albedos, particularly over the oceans and in other remote regions of the world.

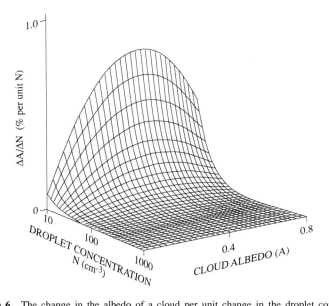

Figure 6   The change in the albedo of a cloud per unit change in the droplet concentration ($\Delta A/\Delta N$) as a function of the cloud albedo (A) and the droplet concentration (N), for a cloud with a constant liquid-water content. (Adapted from Twomey, 1991.)

As discussed in Section I.A.3, clouds that form in air containing high concentrations of CCN tend to have high droplet concentrations, and this enhances the shortwave (i.e., solar) albedo of the clouds. Because clouds are already optically thick at longer wavelengths, changes in the absorption of terrestrial radiation by clouds due to increases in droplet concentration are negligible. Hence, from this point of view, anthropogenic pollution should produce a cooling of the earth's surface. However, there is an opposing effect of aerosol on clouds: increasing anthropogenic emissions of light-absorbing particles (e.g., carbon), which can become incorporated into cloud droplets, will increase the absorption of solar radiation by clouds and therefore tend to warm the earth's surface. Twomey et al. (1984) concluded that on a global scale the "brightening" (i.e., cooling) effect is probably dominant, and that the climatic effect of increasing aerosol concentrations could be comparable (but opposite in sign) to that of global warming due to increasing concentrations of atmospheric carbon dioxide.

Another possible effect of changing atmospheric aerosol concentrations is on cloud extent. As we have seen, increases in cloud droplet concentrations may inhibit precipitation and therefore increase cloud lifetimes. Since this effect should be greater for low clouds, for which the reflection of solar radiation dominates over the absorption of terrestrial radiation, the net effect is expected to be a cooling of the earth's surface. However, no attempts have been made to quantify this effect and there is no observational evidence to suggest that it is occurring.

If aerosol affect clouds, and therefore temperatures on a global scale, feedback mechanisms can be expected to occur. For example, rising temperatures would allow the air to hold more vapor, which would have a direct warming effect on the earth's surface (water vapor is a greenhouse gas), as well as potential effects on cloud liquid-water contents. Also, changes in cloud and precipitation patterns could potentially affect the hydrologic cycle and the energy balance of the earth–atmosphere system. Another possible feedback mechanism is discussed in Section I.B.4. However, we will first describe several examples of the effects of aerosol on the radiative properties of clouds.

2. Effects of Fossil Fuel and Biomass Burning

Sulfur dioxide emissions from anthropogenic sources (primarily fossil fuel burning) have increased significantly over the past century, and the sulfate aerosol that derive from these emissions now dominate aerosol sources in the northern hemisphere, although not in the southern hemisphere (e.g., Cullis and Hirschler, 1980). Since atmospheric sulfates are water soluble and occur in the size range of a few tenths of a micrometer, they are excellent CCN. There is no doubt that on a local, perhaps even regional, scale anthropogenic sources of sulfate affect both CCN concentrations and cloud droplet spectra (some evidence was given in Section I.A.3 above). The question is whether increasing atmospheric sulfate concentrations are having any significant hemispheric global effects. Schwartz (1988) argued that they are not, because he found no evidence for any systematic differences in cloud albedo or surface temperatures between the northern and southern hemispheres, even though anthropogenic emissions of sulfur are much greater in the northern hemisphere. However, as we have seen, it is in clean regions that the albedo of clouds is most susceptible to modification; therefore, the southern hemisphere could be modified by much lower particle emissions than the northern. A recent review by Charlson et al. (1992) suggests that the direct and indirect effects of sulfate aerosol on the global energy balance are of the same sign (cooling) and that their net effect in the northern hemisphere is comparable (but opposite in sign) to current greenhouse gas forcing. Falkowski et al. (1992) concluded from an analysis of sulfate data that, although anthropogenic sulfur emissions enhance cloud albedo adjacent to the east coast of the United States, over the central North Atlantic Ocean an anthropogenic effect is not currently discernible.

Kaufman and Nakajima (1992) used NOAA–AVHRR satellite images taken over the Brazilian Amazon Basin during the biomass burning season to study the effects of smoke aerosol on the properties of low cumulus and stratocumulus clouds. The reflectance at 3.75 μm were studied for tens of thousands of clouds. They showed that the presence of dense smoke (with optical thickness of 2.0) reduced the (remotely sensed) size of droplets in continental clouds from 15 to 9 μm. The cloud reflectance at 0.64 μm was also reduced by 0.03 in the presence of dense smoke. This reduction can be explained by the high initial reflectance of clouds in the visible part of the spectrum and the presence of graphitic carbon. In an analytical paper, Kaufman et al. (1991) reviewed the characteristics of the cooling effect due to aerosol-induced increase in cloud albedo, and they applied Twomey's theory to check whether during fossil fuel or biomass burning the radiative balance favors heating or cooling. Using perturbation analysis (burning of a small amount of fuel) they showed that although coal and oil emit 120 times as many CO_2 molecules as SO_2 molecules, each SO_2 molecule could be 50–1100 times more effective in cooling the atmosphere (through the effects of aerosol on cloud albedo) than a CO_2 molecule is in heating it. Note that this ratio accounts for the large difference in aerosol and CO_2 lifetimes (several days and 7–100 years, respectively) in the atmosphere. Kaufman et al. concluded that the cooling effect of coal and oil burning may presently range from 0.4 to 8 times the heating effect. However, this analysis did not consider the direct effects of aerosol backscattering on solar radiation.

Penner et al. (1991, 1992) calculated the climate impact from biomass burning due to both the direct and indirect (cloud) effects of the smoke. They concluded that both these effects should cause global cooling of similar magnitude, and that the net effect is comparable (but opposite in sign) to global warming by greenhouse gases (however, they neglected the absorption of solar radiation by the aerosol). Clearly, if Charlson et al. (1991, 1992) and Penner et al. (1992) are correct, *decreases* in global temperatures should have been observed over the last century.

However, there are large uncertainties in such calculations, due to both model deficiencies and lack of adequate measurements on the radiative effects of aerosol, particularly their indirect effects on clouds. To properly evaluate the potential for anthropogenic sources of pollution to affect climate, a much better inventory of both the natural and anthropogenic sources of CCN over the oceans and continents is needed. At this juncture, too many theories are chasing too few measurements!

3. Ship Tracks in Clouds

Equations (8) and (11) show that the optical thickness (τ) and the albedo (A) of a cloud are most sensitive to perturbations in the number concentration of cloud droplets when N is small. Hence, thin marine stratiform clouds should be particularly susceptible to modification by CCN. So-called *ship tracks* in clouds (i.e., relatively narrow tracks in marine clouds that appear brighter in satellite imagery—Conover, 1966) provide an excellent demonstration of this effect.

Ship tracks have been attributed to CCN from ships increasing the concentration of cloud droplets and therefore increasing the albedo of marine stratiform clouds. Coakley et al. (1987) showed that ship tracks have a higher reflectivity at 3.7 and 0.63 μm for solar radiation than adjacent noncontaminated clouds. Radke et al. (1989) and King et al. (1993) described a detailed case study of two ships tracks, based on *in situ* and remote sensing measurements. Droplet concentrations in the ship tracks were about $70-100$ cm^{-3} greater than in adjacent cloud regions (Fig. 7). Also, the upwelling intensity of radiation at a wavelength of 0.744 μm was ~ 110 W m^{-2} μm^{-1} sr^{-1} in the ship tracks compared to 40 W m^{-2} μm^{-1} sr^{-1} in the adjacent clouds. Hudson (1991) showed that the effluents from a ship contain significant CCN.

In addition to the effects described above, the ship tracks documented by Radke et al. (1989) also had higher liquid water contents than in the adjacent cloud regions. Albrecht (1989) showed that the liquid water contents of marine stratocumulus clouds were substantially less than adiabatic values, while similarly appearing clouds that developed in admixtures of marine and polluted continental air had, in general, liquid water contents close to adiabatic values. He attributed this difference to the higher droplet concentrations, and lower mean droplet radii, in the cloud admixtures restricting the growth of larger drops by collisions, and hence decreasing the depletion of cloud water by drizzle. A similar effect could be responsible for the relatively high liquid water contents of the ship tracks documented by Radke et al. (1989), since the largest drop present in the ship tracks was 800 μm diameter, whereas, some drops as large as 1200 μm diameter were present in the adjacent cloud unaffected by the ships.

4. The DMS–Cloud–Climate Hypothesis

Over the remote oceans, the main source of sulfate particles and CCN appears to be the oxidation of dimethylsulfide (DMS), which is emitted by phytoplankton

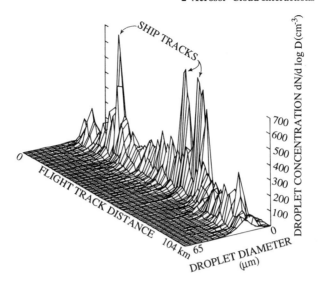

Figure 7   Cloud droplet size spectra (in 3-D perspective) across several ship tracks in a layer of marine stratocumulus cloud. (Data from the University of Washington's Cloud and Aerosol Research Group.)

in seawater. Thus, DMS from the oceans may determine the concentrations and size spectra of cloud droplets, and therefore the cloud albedo, over large regions of the oceans. Of particular importance are marine stratiform clouds, because they cover ~25% of the world's oceans and therefore play an important role in the earth's radiative balance. These ideas are the basis for the DMS–cloud–climate hypothesis (Shaw, 1983; Nguyen et al., 1983; Mészáros, 1988; Charlson et al., 1987).

Figure 8 shows the elements of this hypothesis. DMS from the oceans enters the atmosphere and is subsequently oxidized to sulfate particles, which are assumed to serve as CCN in marine stratiform clouds. If we assume, for example, that DMS emissions increase with increasing temperature (this is not known), then the sequence of processes depicted in Fig. 8 would act as a global thermostat. This is because an increase in DMS emissions would lead to an increase in CCN and cloud droplet concentrations. This, in turn, would increase the amount of solar radiation reflected by marine stratiform cloud, thereby tending to offset the initial increase in global temperature.

An important link in the DMS–cloud–climate hypothesis has recently received some experimental support. This is the assumed dependence of CCN number concentrations in marine air on DMS concentrations. Measurements by Ayers et al. (1991) in Australia, and by Hegg et al. (1991*b*) over the northeastern Pacific

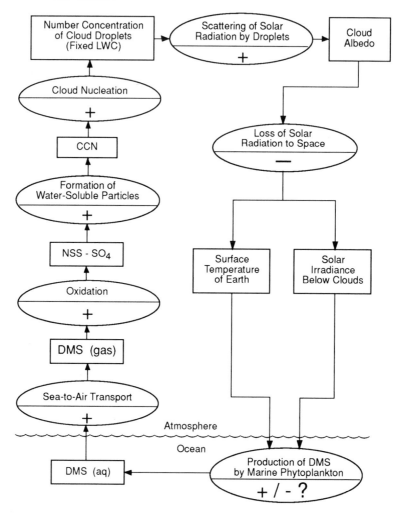

Figure 8   Conceptual diagram of a possible aerosol–cloud climate feedback loop. The + or −
in the ovals indicates the effect of a positive change of the quantity in the preceding rectangle on that
in the succeeding rectangle. (From Charlson et. al., 1987; Copyright © 1987 MacMillan Magazines Ltd.)

Ocean, have shown such a dependence. The Hegg et al. data, shown in Fig. 9, are
represented empirically by

$$[CCN(1.0\%)] = (0.86 \pm 0.25)[DMS] + (23 \pm 17) \qquad (13)$$

and,

$$[CCN(0.3\%)] = (0.28 \pm 0.17)[DMS] + (13 \pm 11) \qquad (14)$$

where [CCN(1%)[and [CCN(0.3%)] are the concentrations (cm^{-3}) of CCN active

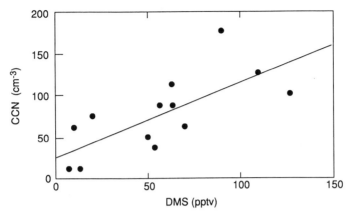

Figure 9 Mean concentrations of CCN active at 1% supersaturation versus concurrent measurements of DMS in the atmospheric boundary layer over the northeast Pacific Ocean. (From Hegg et al., 1991*b*.)

at 1% and 0.3% supersaturations, respectively, and [DMS] is the concentration (pptv) of DMS. From the Ayers et al. data (four points for four seasons) the regression equation is (Hegg et al., 1991*b*):

$$[CCN(1.0\%)] = (0.99 \pm 0.14)[DMS] + (56 \pm 13) \qquad (15)$$

The contrast between the regressions for [CCN(1.0%)] and [CCN(0.3%)] could be important since it is only the larger CCN, active at supersaturations <0.5%, that can affect the microstructure of stratus clouds.

Another important link in the DMS–cloud–climate hypothesis is that between CCN concentrations and cloud droplet concentrations. As discussed in Section I.A.3 and illustrated in Fig. 4, there is experimental evidence that droplet concentrations in marine clouds increase as CCN concentrations rise.

Other important questions remain concerning the DMS–cloud–climate hypothesis. For example, new particles produced by homogeneous nucleation of DMS oxidation products will be on the order of only 0.001 μm (e.g., McMurry and Friedlander, 1979). Particles this small will not be activated by the small supersaturations (≲0.5%) typical of stratiform clouds. As we will see in Section II.B, further processing of these particles by heterogeneous reactions in cumuliform clouds could increase their sizes enough to serve as CCN in stratiform clouds.

C.   Aerosol Effects on Ice in Clouds—Ice Nuclei

The formation of ice in clouds is important because it can quickly lead to precipitation (e.g., Wallace and Hobbs, 1977); the presence of ice also modifies the radiative properties of clouds.

Ice particles can be *homogeneously* nucleated, either directly from the vapor phase or by the freezing of supercooled droplets, at temperatures at and below about $-40°C$. Ice can form at higher temperatures by *heterogeneous* nucleation on a subset of the atmospheric aerosol called *ice nuclei*. Ice nuclei generally satisfy several requirements (e.g., Pruppacher and Klett, 1978). They are insoluble in water; large particles are more effective than small; good ice nuclei generally have chemical bonding and crystallographic structures similar to those of ice; and certain topographic surface features can play an important role in ice nucleation. However, theoretical knowledge of ice nucleation is insufficient to predict with precision the ice nucleating ability of a given material. Some of the more common ice nucleating materials in the air are various biogenic materials, derived from decaying plant leaf litter and the ocean surface (e.g., Schnell and Vali, 1976; Vali et al., 1976), certain clay particles (e.g., Roberts and Hallett, 1968), some combustion products (e.g., Hobbs and Locatelli, 1969), and some pollutants from industries (e.g., Langer, 1968). Some CCN may be latent ice-forming nuclei, which serve first as CCN and then as ice nuclei. For example, Rosinski (1991) observed in the laboratory that the residues from evaporated droplets occasionally served as deposition ice nuclei (i.e., vapor deposited on them directly as ice) at temperatures as high as $-5°C$ and at supersaturations over ice of 0.2%.

Ice nucleus concentrations in the atmosphere are quite variable, both in space and time, but on average they can be represented empirically by

$$\ln X = A(T_1 - T) \qquad (16)$$

where X is the number of ice nuclei per liter of air; T is the air temperature, T_1 is the temperature at which one ice nucleus per liter is active (typically about 253 K), and A is a parameter that varies from about 0.3 to 0.8 deg^{-1} (e.g., Fletcher, 1962). For $A = 0.6$ deg^{-1}, Eq. (16) predicts that X increases by about a factor of 10 for every 4°C decrease in temperature. Since the total concentration of aerosol in polluted air is $\sim 10^8$ per liter, ice nucleus measurements indicate that only about one particle in 10^8 should act as an ice nucleus at $-20°C$. Thus, ice nuclei are much rarer than CCN.

The concentration of ice nuclei is also a function of the supersaturation with respect to ice (S_i) and can be represented empirically by

$$X = BS_i^d \qquad (17)$$

where B and d are constants. Huffman (1973) found $d = 3$ in rural Colorado; $d = 4.5$ in Laramie, Wyoming; and $d = 8$ in St. Louis, Missouri. Note the similarity between Eq. (2), which shows the supersaturation dependence of CCN, and Eq. (17). The reader is referred to Mason (1971), Pruppacher and Klett (1978), and Rogers and DeMott (1991) for further information on the nature, origins, and concentrations of atmospheric ice nuclei.

The equilibrium temperature between ice and an aqueous solution is lower than that between ice and pure water. This is a direct consequence of the lower equilibrium vapor pressure over a solution. Therefore, dissolved materials from CCN (and from chemical reactions within droplets, which will be discussed in Section II.B), will lower the equilibrium melting point of cloud droplets. Similarly, dissolved salts lower the homogeneous and heterogeneous freezing temperatures of droplets.

In addition to the direct action of ice nuclei, *secondary processes* can produce ice particles at temperatures well above the homogeneous nucleation point. (The term *secondary* is something of a misnomer, since these processes can dominate ice particle production in many clouds.) Shown in Fig. 10 are measurements of the maximum concentrations of ice particles (I_M) in marine and continental clouds plotted against cloud-top temperature. The line in Fig. 10 shows the average concentrations of ice particles expected from ice nucleus measurements [Eq. (16)]. It can be seen that while this relation gives an approximate indication of the *minimum* values of I_M, it often underestimates I_M by several orders of magnitude. This is due to the secondary ice-producing processes. The nature of these processes is still a matter of debate (e.g., Rangno and Hobbs, 1991). Possibilities include breakup of primary ice particles produced by nucleation (Hobbs and Farber, 1972), ice splinter production during the freezing of drops (e.g., Mossop, 1985), and enhanced ice nucleating efficiency due to contact nucleation (i.e., an ice nucleus penetrating the surface of a supercooled droplet), or unusually high supersaturations (Hobbs and Rangno, 1985; Rangno and Hobbs, 1991).

A much better predictor of I_M than temperature is the broadness of the cloud droplet size distribution. This is illustrated in Fig. 11, where the same measurements of I_M shown in Fig. 10 are now plotted against a measure of the width of the droplet spectrum, namely, the threshold diameter (defined as the droplet diameter D_T such that the cumulative concentration of droplets with diameters $>D_T$ is 1 cm^{-3}). The relationship between I_M and D_T (the curve in Fig. 11) is

$$I_M = \left(\frac{D_T}{18.5}\right)^{8.4} \qquad (r = 0.90) \qquad (18)$$

As we have seen in Section I.A.3, the CCN spectrum has important influences on the width of cloud droplet spectra and therefore on D_T. Hence, from Eq. (18), CCN spectra can also affect the maximum concentrations of ice particles in clouds.

The growth of ice particles in a cloud by the collection of supercooled droplets, which leads to rimed crystals, graupel, and hail, is affected by the cloud droplet distribution and therefore by CCN. The concentrations and sizes of ice particles in cirrus clouds, which determine radiative scattering by cloud particles, may also

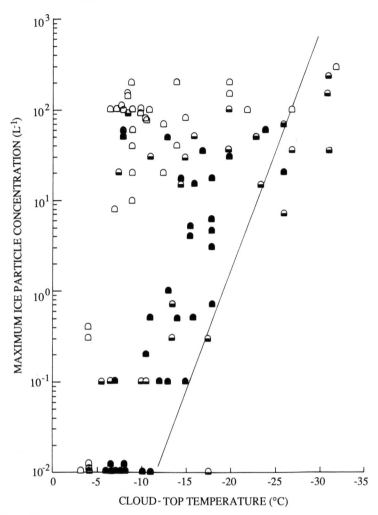

Figure 10 Measurements of maximum ice particle concentrations versus cloud-top temperature in mature and aging marine (open humps), continental (closed humps), and transitional (half-open humps) cumuliform clouds. The line represents the concentrations of ice nuclei given by Eq. (16). (From Hobbs and Rangno, 1985.)

depend on CCN, since many of them are frozen drops (Cooper and Vali, 1981; Sassen and Dodd, 1988; Heymsfield and Sabin, 1989). Because of the large contrast between temperatures of cirrus clouds and the ground, scattering has a significant climatic impact on cirrus longwave radiative properties. Variations in particle size distributions in cirrus clouds could therefore have a large effect on their infrared emissivity.

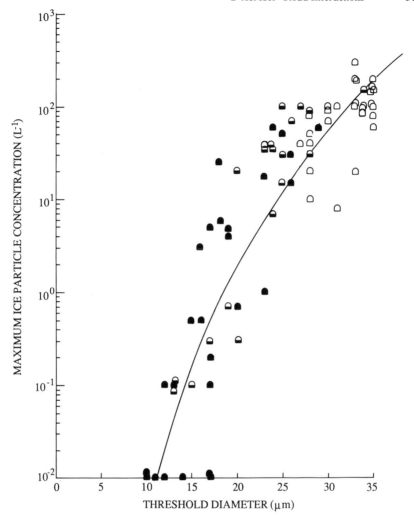

Figure 11   Measurements of maximum ice-particle concentrations versus threshold diameters for mature and aging cumuliform clouds with top temperatures $\leq -6°C$. The curve represents Eq. (18). Symbols are as in Fig. 10. (From Hobbs and Rangno, 1985.)

II.   Effects of Clouds on Aerosol

A.   Scavenging of Aerosol by Clouds

The *scavenging* of aerosol by clouds and their *removal* from the atmosphere by precipitation are important sinks for atmospheric aerosol.[3] The incorporation of

---

[3] For clarity we use here the term *scavenging* to refer to the incorporation of aerosol into cloud droplets, and *removal* to refer to the extraction of aerosol from the atmospheric column. However, this convention is not universally accepted (see Section II.A.2).

aerosols into cloud and precipitation particles also has important effects on the chemical composition of cloud water and precipitation. These topics are discussed in this section.

1. In-Cloud Nucleation Scavenging

In-cloud nucleation scavenging refers to the incorporation of CCN into cloud droplets by activation. The *total scavenging coefficient* (F_s) of aerosol by a cloud can be defined as

$$F_s \equiv \frac{\text{aerosol (ca)} - \text{aerosol (ci)}}{\text{aerosol (ca)}} \tag{19}$$

where "aerosol (ca)" is the concentration of aerosol in the clear air that enters cloud base and "aerosol (ci)" is the concentration of aerosol that is present interstitially between the cloud droplets (or ice particles). Defined in this way, the value of F_s is determined both by nucleation scavenging and by any aerosol mass that is produced within the cloud particles by chemical reactions (see Section II.B).

Junge (1963) predicted sulfate scavenging coefficients due to nucleation scavenging alone in the range 0.5 to 1.0. However, later work showed that they can be higher than this. For example, the relationship between dry particle radius (r_d, in cm) and the supersaturation (S_c, in %) at which a particle will be activated as a CCN is (Fitzgerald, 1973)

$$r_d = 1.53 \times 10^{-6} \, \epsilon^{-0.31} \, S_c^{-2/3} \tag{20}$$

where ϵ is the fraction of water-soluble material in the particle (see also Fitzgerald et al., 1982; Harrison, 1985; Alofs et al., 1989; Okada et al., 1990). Thus, even for $\epsilon = 0.1$, 0.1-μm particles (and thus commonly all of the sulfate mass in the air) will be activated at $S_c = 0.5\%$. Such supersaturations are achieved in clouds with only modest updrafts (0.5–1.0 m s^{-1}).

Scott and Laulainen (1979), considering a single case, compared measurements of cloud-water sulfate concentration and the sulfate concentration in the air in which the cloud formed. They concluded that a nucleation scavenging coefficient of 0.55 could explain the sulfate content of their cloud-water sample, and they estimated that a supersaturation as low as 0.1% could have produced such an efficiency. Radke (1983) and Leaitch et al. (1983) present data suggestive of nucleation scavenging coefficient >0.5 for sulfate. Daum et al. (1983) presented some limited measurements of cloud interstitial and cloud-water sulfate concentrations, which suggested that the bulk of the sulfate mass is incorporated into cloud droplets. The most extensive data set (36 case studies) on aerosol scavenging coefficients in clouds is that of Hegg et al. (1984a) for cumulus, stratus and stratocumulus clouds. In the case of sulfate particles, they obtained values of F_s ranging from 0.1 to 11. Values of $F_s > 1$ (eleven values) indicate that in these cases significant sulfate was produced by chemical reactions in the droplets. Measurements of the nucleation scavenging coefficients for aerosol in the radius range 0.1 to 1.0 μm (primarily sulfate) yielded an average value of 0.7 ± 0.2 (14 values).

From another data set (23 case studies), Hegg and Hobbs (1988) derived a mean value of F_s for sulfate of 0.53 ± 0.19. The individual values were highly variable due, no doubt, to variations in the sulfate size distributions, updraft velocity, and so on. Leaitch et al. (1986) were able to explain the sulfate content of cloud water in Ontario, Canada, by nucleation scavenging of sulfate aerosol in at least 9 out of 11 cases studied. Computations by Flossmann and Pruppacher (1988) for a convective cloud over Hawaii showed that in-cloud scavenging of aerosol was mainly controlled by nucleation scavenging, while impact scavenging played a negligible role (below-cloud impact scavenging—see Section II.A.2—contributed only 5% to the overall particle scavenging and ~40% to the aerosol mass in the rain on the ground). However, SO_2 scavenging was not included in this model. In a similar numerical study for warm clouds over the Atlantic Ocean, Flossmann (1991) found that about 90% of the total amounts of aerosol mass scavenged was incorporated into the cloud water by nucleation scavenging.

Measurements by Hegg et al. (1984a) and Hegg and Hobbs (1986, 1988) on cumulus, stratus, and stratocumulus clouds suggested that nucleation scavenging and/or gas scavenging are major sources of nitrate in cloud water. One data set (23 case studies) yielded a nucleation scavenging coefficient for nitrate of 0.44 ± 0.27 (Hegg and Hobbs, 1988) but, as in the case of sulfate, the individual values were quite variable.

Aerosol ingested into a cloud that do not serve as CCN will form cloud interstitial aerosol (CIA). Both CCN and CIA should change as a cloud ages. The CIA should slowly diminish in number due to coagulation with cloud particles. Twomey (1972) gives an example of a cloud of droplets ~10 μm in diameter in concentrations of 500 cm^{-3}, in which the CIA, consisting of ~0.05 μm diameter particles, were calculated to be removed at a rate of only a few percentages per hour. Since atmospheric aerosol <0.1 μm in diameter generally contain an appreciable soluble component, some of the CIA will be in the haze state. This will reduce coagulation. Thus, in Twomey's example, if the CIA double in size the coagulation rate is more than halved.

Radke (1983) measured CIA in cumulus and stratocumulus clouds. He found that most of the aerosol number and much of the aerosol mass just below cloud base are incorporated into cloud droplets by nucleation scavenging. The number concentration of 0.2 μm diameter CIA was, on average, ~15% of the total concentration just below cloud base, with a somewhat larger percentage for particles <0.5 μm.

In a numerical modeling study, Ahr et al. (1989) found that CIA have a size distribution with a sharp cutoff at specific radii for the dry and wet particle size distributions. Particles above the limiting size acted as CCN. The limiting size depended on the supersaturation, but it was independent of the chemical composition of the particles. Below the limiting size, the CIA spectrum was depleted of dry aerosol in a manner that did depend on their chemical composition and on the supersaturation. The number of aerosol that served as CCN depended critically on their chemical composition and the supersaturation of the air.

2. Below-Cloud Removal of Aerosol by Precipitation

The removal of aerosol by precipitation is one of the major processes by which the atmosphere is cleansed and a balance maintained between the sources and sinks of atmospheric aerosol. It has been estimated that in temperate latitudes precipitation removes about $70-80\%$ of the mass of the aerosol in the tropopause (SMIC, 1971).

The concentration of aerosol $X(D_p)$ with diameters between D_p and $D_p + dD_p$ after removal that has been active for a time period t can be represented by

$$X(D_p) = X_0(D_p) \exp[-\Lambda(D_p)t] \qquad (21)$$

where $X_0(D_p)$ is the aerosol concentration at $t = 0$ and $\Lambda(D_p)$ is a parameter called the *scavenging rate*.[4] The *scavenging collection efficiency* $E(D_p, D)$ of aerosol of diameter D_p by precipitation particles of size D can be derived from

$$\Lambda(D_p) = \int_0^\infty A(D)E(D_p, D)V(D)N(D) \, dD \qquad (22)$$

where $A(D)$ is the effective cross-sectional area of a precipitation particle, $V(D)$ is the fallspeed of the particle, and $N(D) \, dD$ the concentration of precipitation particles with diameters between D and $D + dD$. If it is assumed that E is a function only of D_p, then

$$E \simeq \Lambda(D_p) \left[\sum_i A_i(D)V_i(D)N_i(D) \, \Delta D_i \right]^{-1} \qquad (23)$$

where i indicates a size interval for the precipitation particles.

Aerosol may be removed by precipitation due to inertial (or gravitational) impaction, electrical and phoretic effects, and Brownian motion (see Pruppacher and Klett, 1978, for a review). Theoretical and numerical modeling studies of these removal mechanisms have led to predictions of the functional dependence of the scavenging collection efficiency on aerosol size (e.g., Dana and Hales, 1976; Grover et al., 1977; Wang et al., 1978; Leong and Beard, 1978; Flossmann and Pruppacher, 1988). Shown in Fig. 12a are the results of some of the theoretical calculations and one set of laboratory measurements. Fig. 12b shows some field measurements of E.

Despite the wide variety of aerosol and precipitation particles, the various results in Fig. 12 show the same general features, namely: a minimum in E near $D_p = 1$ μm (called the "scavenging gap"); increasing values of E with increasing

---

[4]The term *removal* would be preferable to *scavenging* (see footnote 3). However, to avoid confusion, we employ here the more widely used term *scavenging*.

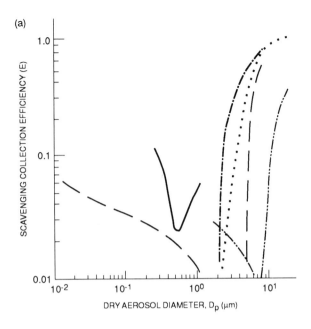

(a)

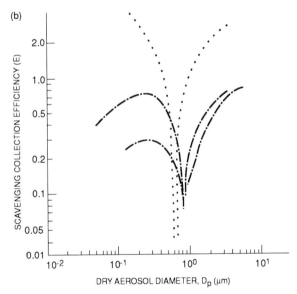

(b)

Figure 12 (a) Precipitation scavenging collection efficiencies predicted theoretically by Wang et al. (1978) for scavenging drops 84 μm (– –) and 620 μm (– · – · –) in diameter, by Leong and Beard (1978) for 80 μm in diameter (– ·· –) and by Dana and Hales (1976) for drops 50 μm in diameter (· · · · ·). The continuous line (————) shows the results from laboratory measurements by Lai et al. (1978) for scavenging drops 620 μm in diameter. (b) Precipitation scavenging collection efficiencies measured in the field for natural aerosol (– · –) and for volcanic particles (· · · · ·). (From Radke et al., 1980.)

D_p for $D_p > 1$ μm (due to the dominant role of inertial impaction for this size range of aerosol); and increasing values of E with decreasing D_p to the left of the scavenging gap, where Brownian and phoretic forces dominate (see Pruppacher and Klett, 1978). The main discrepancy between the field measurements shown in Fig. 12b and the results shown in Fig. 12a is that the values of E in Fig. 12b for submicrometer aerosol are about an order of magnitude larger than the theories predict for scavenging by Brownian and phoretic effects alone. Indeed, outside of the scavenging gap, the field measurements lend support to Slinn's (1974) suggestion that E is close to unity for submicron particles. For a discussion of processes that may be responsible for some of the differences among the various results shown in Fig. 12, the reader is referred to Radke et al. (1980).

A scavenging collection efficiency with the general shape of the curves shown in Fig. 12 should have a strong impact on the size distribution of atmospheric aerosol. Outside of the scavenging gap, aerosol should be efficiently removed by precipitation, but within the scavenging gap removal by precipitation will be very inefficient. Consequently, we would expect to see relatively high concentrations of particles with dry diameters of about 0.5 to 1 μm. This size range corresponds reasonably well to the peak in concentrations observed in atmospheric aerosol (Willke and Whitby, 1975), although this peak is generally ascribed to processes other than the scavenging minimum.

Flossmann and Pruppacher (1988) carried out a theoretical study of the wet removal of atmospheric pollutants by a convective cloud. Some results from this study are shown in Fig. 13. The total aerosol mass in the air decreases very

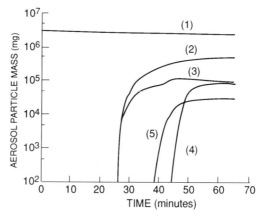

Figure 13 Results from a theoretical study of the scavenging of aerosol particles by a convective cloud. Total aerosol mass in the air (curve 1), cumulative aerosol mass scavenged by nucleation (curve 2), total aerosol mass in cloud water (curve 3), cumulative aerosol mass in rain on ground (curve 4), and cumulative aerosol mass scavenged by impaction scavenging (curve 5), as a function of time. (From Flossmann and Pruppacher, 1988.)

slightly with time as particles are consumed by the drops and as these are subsequently deposited on the ground (curve 1). Nucleation scavenging (curve 2)
begins at the onset of cloud formation. This time correlates with the time when
the aerosol is first found in the cloud water (curve 3). The amount of aerosol mass
removed by impaction scavenging (curve 5) is considerably smaller than that removed by nucleation scavenging. It remains insignificant until precipitation-size
drops start leaving the cloud base. This is a result of the fact that inside the cloud
the main aerosol mass is consumed by the nucleation of drops on the larger
aerosol, leaving only a negligible mass of small interstitial aerosol for impaction
scavenging. The effects of impaction scavenging become noticeable as soon as
raindrops leave cloud base and fall through air where there are still numerous large
aerosol. Aerosol scavenged by both nucleation scavenging and impaction reach
the ground (inside raindrops) shortly thereafter (curve 4). The sum of curves (2)
and (5) give the cumulative scavenged aerosol mass. This sum is larger than the
sum of curves (3) and (4) because some drops evaporated and released the scavenged aerosol back into the air.

Field measurements indicate that snow crystals can contain concentrations of
aerosol up to twice that of raindrops for equivalent precipitation rates (e.g., Englemann and Perkins, 1966; Carnuth, 1967; Sood and Jackson, 1969, 1970; Knutson
et al., 1976). A recent theoretical treatment of this problem by Miller (1990) incorporates the effects of Brownian diffusion, thermophoresis (i.e., migration of
aerosol down a temperature gradient), diffusiophoresis (i.e., migration of aerosol
down a gradient in vapor density), and electrostatic attraction. However, since it
does not include inertial impactions, this theory is applicable only to aerosol particles less than 1 μm radius. Scavenging rates per snow crystal cross-sectional area
are greatest for spheres, less for discs, and least for needles. Increasing temperature increases aerosol mobility and hence increases scavenging by Brownian
diffusion. Increasing temperature also increases scavenging by thermophoresis,
which is most significant for aerosol particles near 1 μm radius. Thermodiffusio-
phoresis decreases rapidly as the relative humidity (RH) approaches 100%. At
saturation only Brownian diffusion and electrical forces act. Variation in RH are
much more significant than variations in crystal type. For example, a change in
RH from 95 to 99% changes the scavenging rate by an order of magnitude, with
higher rates at the lower RH. For unsaturated conditions, the phoretic effect essentially fills in the scavenging gap, the region where Brownian diffusion and inertial
impaction become insignificant.

In mixed clouds, where ice crystals grow by vapor deposition at the expense of
droplets, the diffusiophoretic force drives aerosol toward the ice crystals, but the
thermophoretic force is in the opposite direction (because the latent heat of deposition raises the surface temperature of growing crystals). Young (1974) carried
out a numerical study of the effects of both thermophoresis and diffusiophoresis
for a 10 μm radius drop and found thermophoresis to be larger for aerosol parti-

cles less than 1 μm radius and diffusiophoresis to be larger for aerosol particles greater than 1 μm.

B. Chemical Reactions in Clouds and Their Effects on Aerosol

Radke and Hobbs (1969), Saxena et al. (1970), Dinger et al. (1970), Radke (1970) and Hegg et al. (1980) observed that CCN concentrations active at a given supersaturation are often higher in air that has been processed by clouds than in the ambient air. Radke and Hobbs (1969) and Hobbs (1971) suggested that this enhancement in CCN activity is due to the oxidation of SO_2 to form sulfate in cloud droplets. Thus, when a cloud droplet evaporates it leaves behind both the initial CCN on which it formed and any material formed by chemical reactions in the droplet that precipitates out of solution when the droplet evaporates. If these two sources of materials in each droplet combine to form one particle when the droplet evaporates, the number of particles in the air will not be affected by the formation and evaporation of clouds, but the mass of atmospheric aerosol will be increased. Also, since each original CCN will have been increased in size, the CCN spectrum of the particles released by a cloud will be different (i.e., enhanced in activity) than that ingested by the clouds.

Wave clouds, which form when air flowing over a mountain is lifted above its condensation level, provide a relatively simple system for exploring chemical reactions in clouds and their effects on atmospheric aerosol. In fact, the first interactive cloud-physics–cloud-chemistry model dealt with wave clouds (Easter and Hobbs, 1974). In this model, a parcel of air is followed as it moves along a streamline through a wave cloud. As the parcel enters the cloud, droplets form on existing CCN; these droplets grow and evaporate as the parcel rises and descends. Simultaneously, SO_2 and NH_3 dissolve in the droplets and the dissolved SO_2 is oxidized to sulfate ions, following the chemistry proposed by Scott and Hobbs (1967) and using first-order rate constants covering a range of values determined from earlier laboratory studies. When the air parcel leaves the wave cloud and the droplets evaporate, each CCN on which a droplet formed has associated with it some additional mass of ammonium sulfate [actually a molecular form somewhere between H_2SO_4 and $(NH_4)_2SO_2$] produced during transit of the cloud. The amount of this ammonium sulfate, and the corresponding enhancement of the activity of the CCN, is predicted by the model calculations. Some results are shown in Fig. 14. This shows that a portion of the CCN in the air entering the cloud were activated in the cloud at supersaturations between about 0.6 and 1.4%. However, after leaving the cloud, the additional sulfate deposited on these particles causes them to be activated at supersaturations between about 0.2 and 0.6%. Updates of these model calculations, using more recent estimates of ambient trace gas concentrations and oxidation rates, were made by Hegg and Hobbs (1979, 1981).

Hegg and Hobbs (1982) tested the above idea by measuring the amount of sulfate entering and exiting 28 wave clouds in the Pacific Northwest of the United

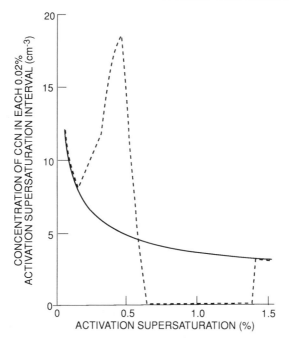

Figure 14 Model calculations of the effects of passage through a wave cloud on CCN activated at a given supersaturation. The solid line is for the air entering the wave cloud and the dashed line is for the air exiting the wave cloud. Oxidation rate after McKay (1971); SO_2 and NH_3 concentrations, 1 and 3 ppb, respectively. (From Easter and Hobbs, 1974.)

States. The measured sulfate production was found to be quite variable—on some occasions it was as high as 10.9 ± 3.8 μg m^{-3} and on other occasions it did not differ significantly from zero. Similar studies of wave clouds by Chandler et al. (1989) and Gallagher et al. (1990) in England have confirmed sulfate production by cloud processing and have helped to identify some of the factors responsible for its variability.

From an analysis of measurements made on 13 cumuliform clouds in the Pacific Northwest, Hegg and Hobbs (1986) derived a mean sulfate production of 0.9 ± 0.5 μg m^{-3}. Hegg and Hobbs (1988) obtained an extensive data set on sulfate production in clouds on the Mid-Atlantic and Pacific Northwest coasts of the United States. The mean sulfate production for the entire data set was 1.99 ± 0.48 μg m^{-3}. However, there was a noticeable difference between the two coasts: for the Mid-Atlantic it was 2.84 ± 0.83 μg m^{-3} and for the Pacific Northwest it was 0.97 ± 0.33 μg m^{-3}. This difference is probably due to higher levels of pollution on the East Coast.

Hegg (1985) concluded that for the troposphere as a whole, the conversion of SO_2 to sulfate in clouds was about 10 to 15 times greater as a sink for SO_2 than

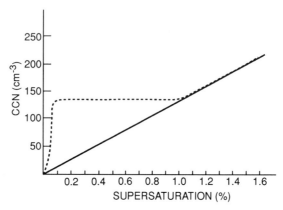

Figure 15 Numerical model predictions of the CCN spectrum left behind by an evaporating cumulus cloud (dashed line) compared to the CCN spectrum that entered the cloud (continuous line). (From Hegg, 1990.)

homogeneous gas-phase oxidation. Hegg (1990) concluded that in the remote marine atmosphere aerosol have to be processed by cumuliform clouds before they will be large enough to serve as CCN at the low supersaturations typical of marine stratus clouds. Some results from Hegg's model calculations are shown in Fig. 15. The CCN spectrum input into the cumulus clouds (the continuous line in Fig. 15) has a CCN number concentration typical of that expected in the marine atmosphere from homogeneous (gas phase) processes. Since the maximum supersaturation achieved in stratiform clouds is generally 0.5% or less, this CCN spectrum would produce only about 50 cm^{-3} of droplets in a stratiform cloud. However, after this CCN spectrum has been processed by a cumuliform cloud, the additional sulfate produced by chemical reactions in the cloud is deposited onto the original CCN when the cloud drops evaporate. The CCN spectrum that emerges is capable of activating about 130 cm^{-3} of droplets in a stratiform cloud (dashed line in Fig. 15).

The oxidants and oxidation mechanisms responsible for sulfate production in clouds have received considerable attention in recent years. Candidate oxidants include hydrogen peroxide (e.g., Penkett et al., 1979; Daum et al., 1983), ozone (e.g., Penkett et al., 1979; Maahs, 1983; Martin, 1984; Hoffman, 1986; Hegg and Hobbs, 1978, 1987), and oxygen (e.g., Hegg and Hobbs, 1978). It appears that all of these oxidants can contribute to sulfate production in clouds; the environmental conditions determine which dominates in any particular situation.

There is also some evidence for nitrate production in clouds. Hegg and Hobbs (1988) derived a value of 1.71 ± 0.68 µg m^{-3} from 23 case studies on the Mid-Atlantic and Pacific Northwest coasts. As in the case of sulfate production, nitrate production for the Mid-Atlantic coast (2.91 ± 1.07 µg m^{-3}) was greater than for the Pacific Northwest coast (0.06 ± 0.10 µg m^{-3}). Leaitch et al. (1986) also obtained evidence for substantial enhancement of nitrate in clouds. However, much more data and theoretical studies are needed on nitrate production in clouds.

C. Acidification of Cloud Water and Precipitation

The processes described above, namely, the nucleation scavenging of aerosol by clouds, the scavenging of aerosol by precipitation, and chemical reactions in cloud droplets, are responsible for the acidification of cloud water and precipitation (Fig. 16).

Hegg (1983) combined field measurements and model calculations to estimate the relative contributions of these three mechanisms to the sulfate content of precipitation from convective clouds. Variations in cloud liquid-water content, and the presence of ice, were found to have substantial impacts on the relative contributions. For all-liquid convective clouds formed in polluted air, the contributions to the sulfate content of the precipitation were ~25% from nucleation scavenging,

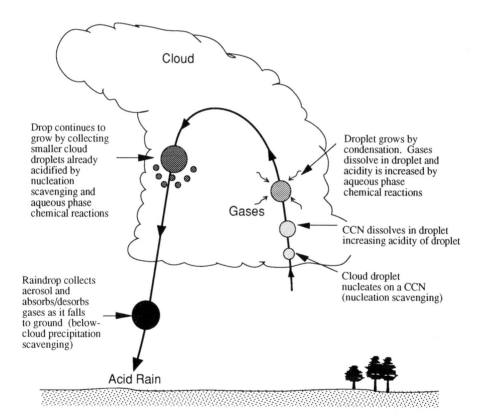

Figure 16   Schematic of processes leading to the acidification of cloud water and precipitation.

~30% from below-cloud impact scavenging, and ~45% from in-cloud chemical reactions. The corresponding percentages for clouds formed in clean marine air were ~65%, ~25%, and ~19%. In the polluted case, the SO_2 concentrations in the ambient air were high enough to permit chemical reactions in clouds to dominate the sulfate content of the precipitation. In the clean marine air, the lower concentrations of sulfur gas reduced the in-cloud chemical reactions, and nucleation scavenging then dominated the sulfate production. Because of the importance of understanding the processes that produce acid precipitation, much numerical modeling work has been done on this subject in recent years. It is beyond the scope of this paper to review these models. The interested reader is referred to the following: for cumulus-scale models, Tremblay and Leighton (1986) and Taylor (1989a,b); for rainbands, Hegg et al. (1984b, 1986, 1989), Hales (1989) and Barth et al. (1992); and for the synoptic scale, Carmichael et al. (1986), Chang et al. (1987), and Venkatram et al. (1988).

D.   Nucleation of Aerosol in and near Clouds

In Section II.B we considered how CCN are modified after being incorporated into, and subsequently released from, cloud droplets. In this section, we describe some recent observations that indicate that new aerosol may be created in the vicinity of, or within, clouds.

    The homogeneous bimolecular nucleation of sulfuric acid droplets from the $H_2O-H_2SO_4$ vapor system has been proposed as an important source of small aerosol particles (i.e., Aitken nuclei) (e.g., Junge, 1963; Kiang et al., 1973; Middleton and Kiang, 1978). However, until recently, little field data were available to support this idea. This may have been due, in part at least, to the fact that there are two possible sinks for supersaturated H_2SO_4 vapor: nucleation of new aerosol, and condensation onto existing aerosol. Under polluted conditions, where the production of H_2SO_4 vapor is large, there is generally a great deal of preexisting aerosol surface on which the vapor can condense. For cleaner conditions, where the surface area of the existing aerosol is much less, precursor gas concentrations (e.g., of SO_2) may be too low for sufficiently rapid production of the acid vapor to achieve the supersaturations necessary for new aerosol production. Thus, the nucleation of new aerosol may occur only under rather special atmospheric conditions. These conditions may exist, on occasions, in the marine atmosphere. Here, the surface area of existing aerosol is often quite low, and SO_2 concentrations are not negligible.

    Recently, Hegg et al. (1990) and Radke and Hobbs (1991) presented data that is strongly suggestive of new aerosol formation in marine air in and around clouds. One such set of measurements shows Aitken nucleus concentrations and humidity measured as a research aircraft flew in and out of five small marine cumulus clouds (Fig. 17). These two parameters were very well correlated, both within and outside of the clouds; the Aitken nucleus concentrations were much

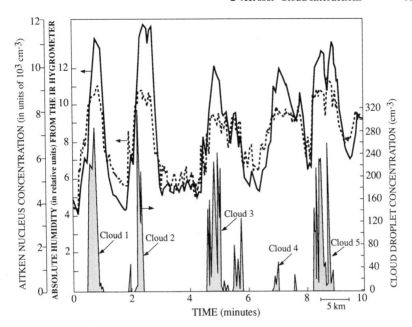

Figure 17   Measurements of Aitken nucleus concentrations (dashed line) and humidity (solid line) across five small marine cumulus clouds (shaded regions), indicated by the cloud droplet concentrations. (From Radke and Hobbs, 1991.)

tions were much higher in the clouds and in "halo" regions surrounding the clouds than they were in the air well removed from the clouds. Hegg et al. (1990) presented similar evidence for aerosol production in marine stratiform clouds.

Hegg et al. (1990) explored the main features of aerosol production in the vicinity of marine clouds in terms of homogeneous-bimolecular theory (e.g., Warren and Seinfeld, 1985; Kreidenweiss and Seinfeld, 1988). This theory is similar to that for the homogeneous nucleation of water droplets from water vapor, except that two molecules are involved in the condensation (sulfuric acid and water molecules, in this case). The source of the sulfuric acid is SO_2, which, as we have seen (Section I.B.4), derives from DMS in marine air. The theory also allows for competition between new aerosol formation and condensation onto existing aerosol.

Hegg et al. (1990) also used this model to explore the influence of various parameters on the nucleation of new aerosol in marine air. At low temperatures, new particle production was found to be favored over condensation onto existing aerosol. New aerosol production increased with increasing SO_2. Below an SO_2 concentration of $25-50$ pptv, the model predicted negligible new aerosol production (for an assumed OH concentration of 10^6 molecules cm^{-3}). Above these SO_2 concentrations, new aerosol production increased rapidly with increasing SO_2. The effect of relative humidity (RH) is shown in Fig. 18. Only when the RH is

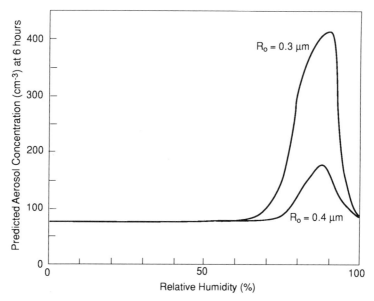

Figure 18   Model predictions of aerosol concentrations, produced by the nucleation of new aerosol in marine air, as a function of initial relative humidity after 6 hours of daylight. Temperature is 0°C, initial SO_2 concentrations 100 pptv and OH concentration 10^6 molecules cm^{-3}. Curves are shown for two values of the preexisting particle mass mean radius (R_0). (From Hegg et al., 1990.)

greater than about 70% does new aerosol production increase significantly. This is why aerosol production occurs in and around clouds, because clouds raise the RH of the ambient air. The aerosol production reaches a peak at RH ≈ 90%. The peak occurs because the nucleation rate for new sulfuric acid aerosol increases monotonically and rapidly as the RH rises above 70%, but the sink for sulfuric acid onto existing aerosol only begins to increase rapidly when the RH rises above 85–90%.

The production of new aerosol by the homogeneous-bimolecular mechanism outlined above is likely to be most effective just outside the boundaries of clouds, rather than within clouds. This is because the aerosol production rate diminishes as the RH increases above about 90% (Fig. 18), and cloud droplets provide a large sink for gas phase H_2SO_4. However, as we have seen, there is some evidence for maximum aerosol production within marine clouds (see Fig. 17, also Saxena and Rathore, 1984, and Hegg et al., 1990). Hegg (1991) has offered the following explanation for a large aerosol production rate within clouds.

Chameides and Davis (1982) showed that very active photochemistry can go on in the interior of clouds. Also, Madronich (1987) showed that the actinic radiation fluxes (i.e., shortwave ultraviolet radiation that can induce photochemical reactions) in clouds may be several times greater in clouds than out of clouds due to multiple scattering by cloud droplets. This raises the possibility that high OH concentrations could be produced within clouds. This, in turn, could produce a gas phase H_2SO_4 supply rate sufficiently fast to compensate for the high droplet surface-area sink for H_2SO_4, and thus produce concentrations of H_2SO_4 sufficiently high to generate a large aerosol production rate. In model calculations, similar to those of Hegg et al. (1990) but including the effects of enhanced actinic flux, Hegg (1991) obtained significant in-cloud production of new aerosol and interstitial aerosol concentrations similar to those measured by Hegg et al. (1990).

Finally, a word of caution. Although observational evidence is mounting that aerosol are produced within and in the vicinity of clouds, and the homogeneous-bimolecular theory provides a viable explanation for such production, the issue is not yet settled. For example, Hudson and Frisbie (1991) found that large aerosol concentrations above stratus clouds off the Southern California coast were associated with high ozone concentrations, and may therefore have had an anthropogenic origin. In view of the potential implications for climate of clouds acting as a source of aerosol, further research is needed to resolve this issue.

III. Conclusions

In this chapter we have reviewed current understanding of the interactions between atmospheric aerosol and clouds. The physical and chemical processes involved in these interactions in marine and continental air are summarized schematically in Fig. 19. A few of the key elements and uncertainties depicted in this figure are worth emphasizing.

The major sources of CCN in marine air derive from the ocean as DMS, methanesulfonic acid (MSA), sea salt, and organics (Fig. 19a). The factors that determine the emissions of these species from the ocean, and their relative contributions to sulfate aerosol, have not been well quantified. It appears that to reach the sizes required to serve as CCN at the low supersaturations typical of marine stratiform clouds, the sulfate aerosol produced by homogeneous, gas-phase reactions must first be processed by cumuliform clouds. The CCN activity of aerosol released from cumuliform clouds is greater than those that enter it because of the addition of material produced by chemical reactions in the cloud droplets. These same aqueous-phase reactions, together with cloud nucleation scavenging of the CCN, are mainly responsible for any acidification of rain over the remote oceans. Aerosol may be produced in and around marine clouds, probably by homogeneous-bimolecular nucleation.

The corresponding aerosol–cloud interactions over the continents are depicted in Fig. 19b. The main sources of sulfur and nitrogen gases are from anthropogenic and biomass combustion. These gases can be absorbed into cloud particles; they are oxidized to form aerosol, some of which can serve as CCN. CCN are also injected directly into the atmosphere from the earth's surface. Aqueous-phase

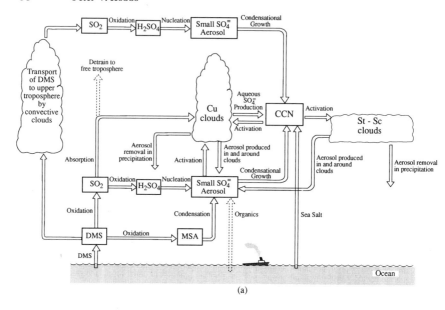

(a)

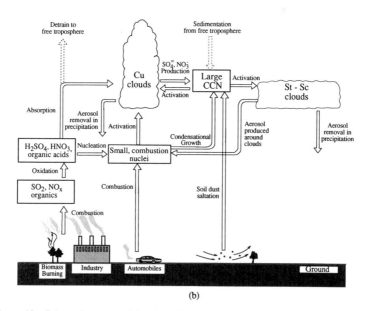

(b)

Figure 19   Schematics summarizing the principal aerosol–cloud interactions in (a) marine air, and (b) continental air.

chemical reactions in clouds can enhance the activity of CCN released from evaporating clouds. Also, in-cloud chemical reactions are probably often the main mechanism for acidifying cloud water and precipitation in polluted air, with nucleation scavenging and below-cloud removal generally playing important but lesser roles. New aerosol production in and around continental clouds is less likely than for marine clouds (and has not been observed to date) because of the higher concentrations of existing aerosol and cloud droplets (which serve as large sinks for condensing gases) in continental air.

A number of issues concerning aerosol–cloud interactions remain uncertain and therefore require further investigation. A global inventory of the principal sources and nature of CCN is required. For example, the role of organics in CCN formation is potentially important but has not been quantified. More information is needed on CCN spectra, particularly long time-series measurements at a variety of locations around the world. The role of clouds as a source of aerosol and CCN is potentially of global importance, but more data are needed to determine the magnitude of this source and the mechanisms involved in the production of the aerosol. The potential effects of aerosol on cloud structures and radiative properties are well demonstrated by ship tracks in clouds. However, the regional and global effects of aerosol on cloud radiative properties remain uncertain. Finally, as observational and theoretical studies establish a broad data base and a firm understanding of the relevant physical and chemical processes, increasingly sophisticated aerosol–cloud interactions should be incorporated into numerical models of precipitation, acid deposition, and climate on local, regional, and global scales.

Acknowledgments

This paper is inscribed to Lawrence F. Radke and Dean A. Hegg, two seekers after truth, with whom I have had the pleasure to work on aerosol–cloud interactions, and many other research topics, for more than 20 years. Preparation of this paper was supported in part by the Quest for Truth Foundation.

References

Ackerman, A. S., Toon, O. B., and Hobbs, P. V. (1992). Presented at *Intern. Conf. on Clouds and Precipitation,* Montreal, 17–21 August 1992.
Ahr, M., Flossman, A. I., and Pruppacher, H. R. (1989). *J. Atmos. Sci.* **9**, 465–478.
Aitken, J. (1923). "Collected Scientific Papers" (E. G. Knott, ed.) Cambridge Univ. Press.
Albrecht, B. A. (1989). Science **245**, 1227–1230.
Alofs, D. J., Hagen, D. E., and Trueblood, M. B. (1989). *J. Appl. Meteor.* **28**, 126–136.
Ayers, G. P., Ivey, J. P., and Gillett, R. W. (1991). *Nature* **349**, 404–406.
Baker, M. B., and Charlson, R. J. (1990). *Nature* **345**, 142–144.
Barth, M. C., Hegg, D.A., and Hobbs, P.V. (1992). *J. Geophys. Res.* **97**, 5825–5845.

Barth, M. C., Hegg, D.A., and Hobbs, P.V. (1992). *J. Geophys. Res.* **97**, 5825–5845.

Bigg, E. L. (1990). *Atmos. Res.* **25**, 583–600.

Bower, K. N., and Choularton, T. W. (1988). *Q. J. Roy. Meteor. Soc.* **114**, 1411–1434.

Carmichael, G. R., Peters, L. K., and Kitada, T. (1986). *Atmos. Environ.* **20**, 173–188.

Carnuth, W. (1967). *Arch. Meteor. Geophys. Bioklimatol.* **A6**, 321–343.

Chameides, W. L., and Davis, D. D. (1982). *J. Geophys. Res.* **87**, 4863–4877.

Chandler, A. S., Choularton, T. W., Dollard, G. J., Gay, M. J., Gallagher, M. W., Hill, T. A., Jones, B. M. R., Penkett, S. A., Tyler, B. J., and Bandy, B. (1989). *Q. J. Roy. Meteor. Soc.* **115**, 397–420.

Chang, J. S., Brost, R. A., Isaksen, I. S. A., Madronich, S., Middleton, P., Stockwell, W. R., and Walcek, C. J. (1987). *J. Geophys. Res.* **92**, 14,681–14,700.

Charlson, R. J., Lovelock, J. E., Andreae, M. O., and Warren, S. G. (1987). *Nature* **326**, 655–661.

Charlson, R. J., Langner, J., Rodhe, H., Leovy, C. B., and Warren, S.G. (1991). *Tellus* **43AB**, 152–163.

Charlson, R. J., Schwartz, S. E., Hales, J. M., Cess, R. D., Coakley, J. A. Jr., Hansen, J. E., and Hofmann, D. J. (1992). *Science* **255**, 423–430.

Coakley, J. A. Jr., Bernstein, R. L., and Durkee, P. A. (1987). *Science* **237**, 1020–1022.

Conover, J. H. (1966). *J. Atmos. Sci.* **23**, 778–785.

Cooper, W. A., and Vali, G. (1981). *J. Atmos. Sci.* **38**, 1244–1254.

Coulier, M. (1875). *J. Pharm. Chim.* **22**, 165–167.

Cullis, C. F., and Hirschler, M. M. (1980). *Atmos. Environ.* **14**, 1263–1278.

Dana, M. R., and Hales, J. M. (1976). *Atmos. Environ.* **10**, 45–50.

Daum, P. H., Schwartz, S. E., and Newman, L. (1983). *In* "Precipitation Scavenging, Dry Deposition, and Resuspension" (H. R. Pruppacher, R. G. Semonin, W. G. N. Slinn, eds.) pp. 31–52. Elsevier Science Pub. Co.

Dinger, J. E., Howell, H. B., and Wojciechowski, T. A. (1970). *J. Atmos. Sci.* **27**, 791–797.

Eagan, R. C., Hobbs, P. V., and Radke, L. F. (1974). *J. Appl. Meteor.* **13**, 553–557.

Easter, R. C., and Hobbs, P. V. (1974). *J. Atmos. Sci.* **31**, 1586–1594.

Englemann, R. J., and Perkins, R. W., (1966). *Nature* **211**, 61–62.

Falkowski, P. G., Kim, Y., Koller, Z., Wilson, C., Wirick, C. and Cess, R. (1992). *Science* **256**, 1311–1313.

Fletcher, N. H. (1962). "The Physics of Rainclouds" Cambridge University Press.

Fitzgerald, J. W. (1973). *J. Atmos. Sci.* **30**, 628–634.

Fitzgerald, J. W., Hoppel, W. A., and Vietti, M. A. (1982). *J. Atmos. Sci.* **39**, 1838–1852.

Flossmann, A. I. (1991). *Tellus* **43B**, 301–321.

Flossmann, A. I., and Pruppacher, H. R. (1988). *J. Atmos. Sci.* **45**, 1857–1871.

Gallagher, M. W., Downer, R. M., Choularton, T.W., Gay, M. J., Stromberg, I., Hill, C. S., Rodojevic, M., Tyler, B. J., Bardy, B. J., Penkett, S. A., Davies, T. J., Dollard, G. J., and Jones, B. M. R. (1990). *J. Geophys. Res.* **95**, 18,517–18,537.

Grover, S. N., Pruppacher, H. R., and Hamielec, A. E. (1977). *J. Atmos. Sci.* **34**, 1655–1663.

Hales, J. M. (1989). *Atmos. Environ.* **23**, 2017–2031.

Harrison, L. (1985). *J. Clim. and Appl. Meteor.* **24**, 312–321.

Hegg, D. A. (1983). *J. Geophys. Res.* **88**, 1369–1374.

Hegg, D. A. (1985). *J. Geophys. Res.* **90**, 3773–3779.

Hegg, D. A. (1986). *J. Atmos. Sci.*, **43**, 399–400.

Hegg, D. A. (1990). *Geophys. Res. Lett.* **17**, 2165–2168.

Hegg, D. A. (1991).*Geophys. Res. Lett.* **18**, 995–998.

Hegg, D. A., and Hobbs, P. V. (1978). *Atmos. Environ.* **12**, 241–254.

Hegg, D. A., and Hobbs, P. V. (1979). *Atmos. Environ.* **13**, 981–987.

Hegg, D. A., and Hobbs, P. V. (1981). *Atmos. Environ.* **15**, 1597–1604.

Hegg, D. A., and Hobbs, P. V. (1982). *Atmos. Environ.* **16**, 2663–2668.

Hegg, D. A., and Hobbs, P. V. (1986). *Atmos. Environ.* **20**, 901–909.

Hegg, D. A., and Hobbs, P. V. (1987). *Geophys. Res. Lett.* **14**, 719–721.

Hegg, D. A., and Hobbs, P. V. (1992). *Proc. 13th Intern. Conf. on Nucleation and Atmospheric Aerosols*, Salt Lake City.

Hegg, D. A., Hobbs, P. V., and Radke, L. F. (1980). *J. Rech. Atmos.* **14,** 217–222.

Hegg, D. A., Hobbs, P. V., and Radke, L. F. (1984*a*). *Atmos. Environ.* **18,** 1939–1946.

Hegg, D. A., Rutledge, S. A., and Hobbs, P. V. (1984*b*). *J. Geophys. Res.* **89,** 7133–7147.

Hegg, D. A., Rutledge, S. A., and Hobbs, P. V. (1986). *J. Geophys. Res.* **91,** 14,403–14,416.

Hegg, D. A., Rutledge, S. A., Hobbs, P. V., Barth, M. C., and Hertzmann, O. (1989). *Q. J. Roy. Meteor. Soc.* **115,** 867–886.

Hegg, D. A., Radke, L. F., and Hobbs, P. V. (1990). *J. Geophys. Res.* **95,** 13,917–13,926.

Hegg, D. A., Radke, L. F., and Hobbs, P. V. (1991*a*). *J. Geophys. Res.* **96,** 18,727–18,733.

Hegg, D. A., Ferek, R. J., Hobbs, P. V., and Radke, L. F. (1991*b*). *J. Geophys. Res.* **96,** 13,189–13,191.

Heymsfield, A. J. and Sabin, R. M. (1989). *J. Atmos. Sci.* **46,** 2252–2264.

Hindman, E. E., Hobbs, P. V., and Radke, L. F. (1977*a*). *J. Air Poll. Contr. Assoc.* **27,** 224–229.

Hindman, E. E., Hobbs, P. V., and Radke, L. F. (1977*b*). *J. Appl. Meteor.* **16,** 745–752.

Hobbs, P. V. (1971). *Q. J. Roy. Meteor. Soc.* **97,** 263–271.

Hobbs, P. V., and Farber, R. (1972). *J. Rech. Atmos.* **6,** 245–258.

Hobbs, P. V., and Locatelli, J. D. (1969). *J. Appl. Meteor.* **8,** 833–834.

Hobbs, P. V., and Rangno, A. L. (1985). *J. Atmos. Sci.* **42,** 2523–2549.

Hobbs, P. V., and Rangno, A. L. (1990*a*). *J. Atmos. Sci.* **47,** 2710–2722.

Hobbs, P. V., and Rangno, A. L. (1990*b*). "Summary of Airborne Data Collected from the University of Washington's Convair C-131A Research Aircraft in Maritime Cumuliform Clouds Off and Near the Washington Coast Between 30 January 1987 and 14 March 1990." Research Report, Cloud and Aerosol Research Group, Atmospheric Sciences Dept., University of Washington.

Hobbs, P. V., and Rangno, A. L. (1992). "Summary of Airborne Data Collected from the University of Washington's Convair C-131A Research Aircraft in Continental and Semi-Continental Cumuliform Clouds Off and Near the Washington Coast Between 27 April 1989 and 12 March 1992." Research Report, Cloud and Aerosol Research Group, Atmospheric Sciences Dept., University of Washington.

Hobbs, P. V., Radke, L. F., and Shumway, S. E. (1970). *J. Atmos. Sci.* **27,** 1216–1217.

Hoffman, M. R. (1986). *Atmos. Environ.* **20,** 1145–1154.

Howell, W. E. (1949). *J. Meteor.* **6,** 134–151.

Hudson, J. G., (1991). *Atmos. Environ.* **25A,** 2449–2455.

Hudson, J. G., and Clarke, A. D. (1992). *J. Geophys. Res.* **97,** 14533–14536.

Hudson, J. G., and Frisbie, P. R. (1991). *J. Geophys. Res.* **96,** 20,795–20,808.

Huffman, P. J. (1973). *J. Appl. Meteor.* **12,** 1080–1082.

Jensen, J. B., and Charlson, R. J. (1984). *Tellus* **36B,** 367–375.

Junge, C. E. (1963). "Air Chemistry and Radioactivity" Academic Press.

Kaufman, Y. J., and Nakajima, T. (1992). *J. Appl. Meteor.* (in press).

Kaufman, Y. J., Fraser, R. S., and Mahomey, R. L. (1991). *J. Climate.* **4,** 578–588.

Kiang, C. S., Stauffer, D., Mohnen, V. A., Bricard, J., and Vigla, D. (1973). *Atmos. Environ.* **7,** 1279–1283.

King, M. D., Radke, L. F., and Hobbs, P. V. (1993). *J. Geophys. Res.* **98,** 2729–2739.

Knutson, E. O., Sood, S. K., and Jackson, M. R. (1976). *Atmos. Environ.* **10,** 395–402.

Köhler, H. (1921). *Geofys. Publ.* 2, No. 3 and No. 6.

Kreidenweiss, S. M., and Seinfeld, J. H. (1988). *Atmos. Environ.* **22,** 283–296.

Lai, K-Y., Dayan, N., and Kerker, M. (1978). *J. Atmos. Sci.* **35,** 674–682.

Langer, G. (1968). *Proc. Conf. on Weather Modifications*, Amer. Meteor. Soc., 220–222.

Leaitch, W. R., Strapp, J. W., Wiebe, H. A., and Isaac, G. A. (1983). *In* "Precipitation Scavenging, Dry Deposition, and Resuspension" (H. R. Pruppacher, R. G. Semonin, and W. G. N. Slinn, eds.), 53–69, Elsevier Science Pub. Co.

Leaitch, W. R., Strapp, J. W., Wiebe, H. A., Anlauf, K. G., and Isaac, G. A. (1986). *J. Geophys. Res.*

Leaitch, W. R., Strapp, J. W., Wiebe, H. A., Anlauf, K. G., and Isaac, G. A. (1986). *J. Geophys. Res.* **91**, 11,821–11,831.

Leong, K. H., and Beard, K. V. (1978). *Preprints Conf. Clouds Physics and Atmospheric Electricity,* Amer. Meteor. Soc., 49–52.

Maahs, H. G. (1983). *Atmos. Environ.* **17**, 341–346.

Madronich, S. (1987). *J. Geophys. Res.* **92**, 9740–9752.

Martin, L.R. (1984). *In* "Oxidation Mechanisms: Atmospheric Considerations" (J. G. Calvert, ed.), Butterworth Publishers, 200 pp.

Mason, B. J. (1971). "The Physics of Clouds" Oxford University Press.

McKay, H. A. C. (1971). *Atmos. Environ.* **5**, 7–14.

McMurry, P. H., and Friedlander, S. K. (1979). *Atmos. Environ.* **13**, 1635–1651.

Meador, W. E., and Weaver, W. R. (1980). *J. Atmos. Sci.* **37**, 630–643.

Mészáros, E. (1988). *Atmos. Environ.* **22**, 423–424.

Middleton, P., and Kiang, C. S. (1978). *J. Aerosol. Sci.* **9**, 359–385.

Miller, N. L. (1990). *Atmos. Res.* **25**, 317–330.

Mossop (1985). *Bull. Amer. Meteor. Soc.* **66**, 264–273.

Nguyen, B.C., Bonsong, B., and Gaudry, A. (1983). *J. Geophys. Res.* **88**, 10,903–10,914.

Okada, K., Tanaka, T., Naruse, H., and Yoshikawa, T. (1990). *Tellus* **42B**, 463–480.

Penkett, S. A., Jones, B. M., Brie, A., and Eggleton, A. E. (1979). *Atmos. Environ.* **13**, 123–137.

Penner, J. E., Dickinson, R. E., and O'Neill, C. A. (1992). *Science* **256**, 1432–1434.

Penner, J. E., Ghan, S. J., and Walton, J. J. (1991). *In* "Global Biomass Burning: Atmospheric, Climate, and Biospheric Implications" (J. S. Levine, ed.), 387–393. The MIT Press, Cambridge.

Pruppacher, H. R., and Klett, J. D. (1978). "Microphysics of Clouds and Precipitation" Reidel.

Radke, L. F. (1970). *Preprints of Papers Presented at Amer. Meteor. Soc. Conf. on Clouds Physics,* Fort Collins, Colorado, 7–8.

Radke, L. F. (1983). *In* "Precipitation Scavenging, Dry Deposition, and Resuspension" (H. R. Pruppacher, R. G. Semonin, and W. G. N. Slinn, eds.), 71–78, Elsevier Science Pub. Co.

Radke, L. F., and Hobbs, P. V. (1969). *J. Atmos. Sci.* **26**, 281–288.

Radke, L. F., and Hobbs, P. V. (1991). *J. Atmos. Sci.* **48**, 1190–1193.

Radke, L. F., Coakley, J. A. Jr., and King, M. D. (1989). *Science* **246**, 1146–1149.

Radke, L. F., Hobbs, P. V., and Eltgroth, M. W. (1980). *J. Appl. Meteor.* **19**, 715–722.

Rangno, A. L., and Hobbs, P. V. (1991). *Q. J. Roy. Meteor. Soc.* **117**, 207–241.

Roberts, P., and Hallett, J. (1968). *Q. J. Roy. Meteor. Soc.* **94**, 25–34.

Rogers, D. C., and DeMott, P. J. (1991). *Rev. Geophys.* (Supplement), 80–87.

Rosinski, J. (1991). *Atmos. Res.* **26**, 509–523.

Sassen, K., and Dodd, G. C. (1988). *J. Atmos. Sci.* **45**, 1357–1369.

Saxena, V. K., Burford, J. N., and Kassner, J. L. (1970). *J. Atmos. Sci.* **27**, 73–80.

Saxena, V. K., and Rathore, R. S. (1984). In *Preprint Volume, 11th International Conf. on Atmospheric Aerosols, Condensation and Ice Nuclei,* Budapest, Hungary, 292–298.

Schnell, R. C., and Vali, G. (1976). *J. Atmos. Sci.* **33**, 1554–1564.

Schwartz, S. E. (1988). *Nature* **336**, 441–443.

Scott, W. D., and Hobbs, P. V. (1967). *J. Atmos. Sci.* **24**, 54–57.

Scott, W. D., and Laulainen, N. S. (1979). *J. Appl. Meteor.* **18**, 138–147.

Shaw, G. (1983). *J. Atmos. Sci.* **40**, 1313–1320.

Slinn, W. C. N. (1974) *In* "Precipitation Scavenging 1974" (R. G. Semonin and A. W. Beadle, eds.). ERDA Symp. Series, 1–60.

SMIC (1971). "Report on the Study of Man's Impact on Climate" MIT Press.

Sood, S. K., and Jackson, M. R. (1969). *Proc. 7th Int. Conf. on Condensation and Ice Nuclei,* Prague and Vienna, 299–303.

Sood, S. K., and Jackson, M. R. (1970). "Precipitation Scavenging." AEC Symposium, Richland, Wash. 121–136.

Squires, P. (1952). *Aust. J. Sci. Res.* **A5**, 59–62 and 473–482.
Squires, P. (1956). *Tellus* **8**, 443–444.
Squires, P. (1958*a*). *Tellus* **10**, 256–261.
Squires, P. (1958*b*). *Tellus* **10**, 262–271.
Taylor, G. R. (1989*a*). *J. Atmos. Sci.* **46**, 1971–1990.
Taylor, G. R. (1989*b*). *J. Atmos. Sci.* **46**, 1991–2007.
Tremblay, A., and Leighton, H. (1986). *J. Clim. Appl. Meteor.* **25**, 652–671.
Twomey, S. (1959). *Geofis Pur. Appl.* **43**, 227–250.
Twomey, S. (1972). *J. Atmos. Sci.* **29**, 1156–1159.
Twomey, S. (1977). "Atmospheric Aerosols" Elsevier.
Twomey, S. (1991). *Atmos. Environ.* **25A**, 2435–2442.
Twomey, S., and Wojciechowski, T. A. (1969). *J. Atmos. Sci.* **26**, 684–688.
Twomey, S. A., Piepgrass, M., and Wolfe, T. L. (1984). *Tellus* **36B**, 356–366.
Vali, G., Christensen, M., Fresh, R. W., Galyen, E. L., Malu, L. R., and Schnell, R. C. (1976). *J. Atmos. Sci.* **33**, 1565–1570.
Venkatram, A., Karamchandani, P. K., and Misra, P. K. (1988). *Atmos. Environ.* **22**, 737–474.
Wallace, J. M., and Hobbs, P. V. (1977). "Atmospheric Sciences: An Introductory Survey" Academic Press.
Wang, P. K., Grover, S. N., and Pruppacher, H. R. (1978). *J. Atmos. Sci.* **35**, 1735–1743.
Warren, D. R., and Seinfeld, J. H. (1985). *J. Colloid Interface Sci.* **105**, 136–143.
Willeke, K. W., and Whitby, K. T. (1975). *J. Air. Pollut. Control Assoc.* **25**, 529–534.
Young, K. C. (1974). *J. Atmos. Sci.* **31**, 768–776.

Chapter 3 | Aerosol–Climate Interactions

Harshvardhan
Department of Earth and Atmospheric Sciences
Purdue University
West Lafayette, Indiana

Particulate matter of submicrometer size in the earth's atmosphere perturbs the radiation field sufficiently to warrant its consideration in any discussion of processes that maintain the current climate. The anthropogenic component of this aerosol burden causes a direct negative (i.e., cooling) radiative forcing of about 1.0 to 2.0 W m^{-2} in the shortwave, which is comparable but of opposite sign to the positive longwave forcing of several trace ("greenhouse") gases associated with industrial and agricultural activities. However, this shortwave forcing is roughly equal to that which would result from an absolute change in global cloud cover of only 1–2%. Therefore, global monitoring efforts and attempts at incorporating aerosols into climate models will invariably be hampered by clouds.

Most atmospheric aerosols of anthropogenic origin are sulfates or smoke, which have residence times in the troposphere of about a week; in contrast greenhouse gases have lifetimes of decades to centuries. A reduction in the emission of sulfur-containing gases, which are the primary precursors of sulfate aerosols, would result in an immediate concomitant decrease in the negative shortwave radiative forcing of the climate system. However, the suggestion that efforts to reduce sulfur emissions could prove harmful by exacerbating greenhouse warming is mischievous considering other deleterious effects of these emissions. Nevertheless, there is some merit in the suggestion that the negative radiative forcing of aerosols may have largely masked to date the global effects of positive greenhouse gas forcing.

The anthropogenic aerosol burden is also being increased by the deliberate large-scale burning of biomass. The smoke produced by these activities results in primarily scattering aerosols that produce an added shortwave radiative forcing. Again, the threat to biodiversity, alterations in surface hydrology, and the possibility of unanticipated climatic consequences resulting from large-scale biomass burning are probably more consequential than any ameliorating climatic effects of these aerosols.

I. Introduction

Over the past decade there has been significant interest in the possibility that long-lived, optically thick, aerosol layers may have modified the earth's climate in the past. Geologic evidence suggests that there have been episodic injections of massive amounts of material into the earth's atmosphere as a result of the impact of large asteroids or comets. The diminution of solar radiation reaching the surface has been cited as the most likely cause of mass extinctions of species (Alvarez et al., 1980; Claeys et al., 1992). The specter of a similar climatic catastrophe following a nuclear war was raised by Crutzen and Birks (1982). Current interest is focused on much more modest injections of materials that form thin aerosol layers in the troposphere. Although the radiative effects are smaller and have been generally ignored in climate models (Hansen and Lacis, 1990), recent studies have estimated that they are not negligible and may be comparable (but opposite in sign) to the radiative effects of increased greenhouse gas emissions (Wigley, 1991; Charlson et al., 1992; Penner et al., 1992).

A. Radiative Forcing

Aerosol layers are composed of submicrometer-sized particles that can be treated independently of the larger particles in clouds in a radiative budget of the earth's atmosphere. The nature of aerosols and their interaction with clouds was covered in the first two chapters of this monograph. This chapter is devoted to the direct radiative effects of aerosols. Furthermore, since there is a vast disparity between the residence times of tropospheric and stratospheric aerosols, there is a natural division when climatic effects of these aerosol layers are studied. Only tropospheric aerosols are considered here; a discussion of stratospheric aerosols appears in Chapter 8.

   The place of direct radiative forcing by tropospheric aerosols in the overall scheme of aerosol–climate interactions is shown in Fig. 1. Any estimates of direct radiative forcing by aerosols can only be translated into a climate response by accepting the verisimilitude of current climate models. The relationship between direct radiative forcing and climate feedbacks shown in Fig. 1 ultimately determines the role aerosols play in the maintenance of the current radiation balance.

   On a global annual mean basis there is a balance between the solar radiation absorbed by the earth–atmosphere system and the longwave emission to space. Changes in any radiatively active component of the system result in perturbations that lead to responses by the climate system as it seeks a new state of equilibrium. The radiative forcing in watts per square meter ($W\ m^{-2}$) is the initial change in absorbed or emitted energy resulting from the prescribed process, such as a change in atmospheric composition, cloudiness, or surface reflectance. It is generally expressed as a change in the net radiative flux at the tropopause. Shine et al. (1990) have tabulated the forcing due to the additional increase in various green-

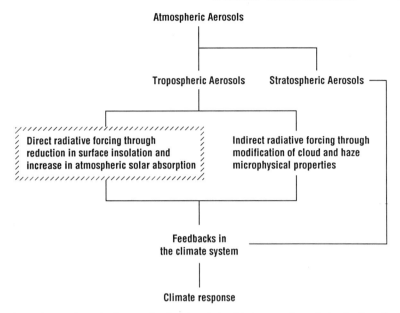

Figure 1   A schematic diagram showing the relationship between the radiative forcing of atmospheric aerosols and climate response.

house gases over preindustrial levels. Their estimate is 1.50 W m^{-2} for carbon dioxide and a cumulative 2.45 W m^{-2} when other greenhouse gases are included. This is a positive forcing in the sense that it results in higher global temperatures. The corresponding forcing for aerosols of natural and various anthropogenic origins can be made by first calculating the radiative properties associated with estimates of aerosol burdens and then computing the change in shortwave and thermal radiation flux produced in the surface–troposphere system.

II. Theoretical Aspects

Some basic aspects of scattering and absorption by small particles typically present in aerosol layers govern the sign and magnitude of the radiative forcing. Radiative properties such as reflectance are chiefly determined by the optical depth, single scattering albedo, and some measure of the scattering phase function. At visible wavelengths, the mean optical depth of tropospheric aerosols ranges from less than 0.05 in remote, pristine environments to about 1.0 near the source of copious emissions such as a forest fire or industrial activity. The optical depth decreases quite rapidly with increasing wavelength if the layer contains only small particles; aerosol layers tend to be fairly transparent at thermal wavelengths. The primary radiative forcing is therefore confined to solar wavelengths. Since there are strong water-vapor absorption bands in the solar near-infrared, the dominant effect of tropospheric aerosols is in the window regions of the solar spectrum, chiefly at visible wavelengths.

A first-order estimate of the radiative forcing can be made by calculating the change in planetary albedo resulting from the addition of the aerosol layer uniformly over the globe. The reflectance R_a of an optically thin layer for incident isotropically diffuse radiation is (Coakley and Chylek, 1975; King and Harshvardhan, 1986)

$$R_a = 2 \varpi \beta \tau = 2 \beta \tau_s \tag{1}$$

and the absorptance A_a is

$$A_a = 2(1 - \varpi)\tau = 2\tau_a \tag{2}$$

In Eqs. (1) and (2), the extinction optical thickness is

$$\tau = \tau_a + \tau_s \tag{3}$$

where τ_a and τ_s are the absorption and scattering optical depths, respectively. The single scattering albedo is

$$\varpi = \tau_s/\tau \tag{4}$$

and β is the average backscatter fraction, which is about 0.3 for aerosol layers. The change in the albedo of the aerosol–surface system ΔR_{as} when the layer is placed over a surface of reflectance R_s, taking into account multiple reflections, can be approximated by

$$\Delta R_{as} \approx R_a(1 - R_s)^2 - 2A_aR_s \tag{5}$$

when $R_a \ll 1$. Substitution of Eqs. (1) and (2) into Eq. (5) leads to the often cited criterion (Chylek and Coakley, 1974) for the albedo perturbation to change sign, which is that the addition of an aerosol layer will increase the system albedo unless

$$(1 - \varpi)/\varpi\beta > (1 - R_s)^2/2R_s \tag{6}$$

The reflection and absorption properties of the layer depend critically on the single scattering albedo ϖ, in addition to the optical depths, which are column integrals of the respective specific extinction coefficients. At visible wavelengths, all constituents of tropospheric aerosols with the exception of elemental carbon are nonabsorbing and $\varpi = 1.0$ (Bohren and Huffman, 1983). Absorbing aerosols contain externally or internally mixed carbon particles; it is general practice to model this mixture as an aerosol of effective absorption properties related in some manner to the carbon content (Ackerman and Toon, 1981). Extinction properties depend on the size of the particle relative to the wavelength and the complex index of refraction, $m = n - ik$, with the imaginary index being a measure of particle absorption. Figure 2a shows the computed values of ϖ at a wavelength of 0.63 μm

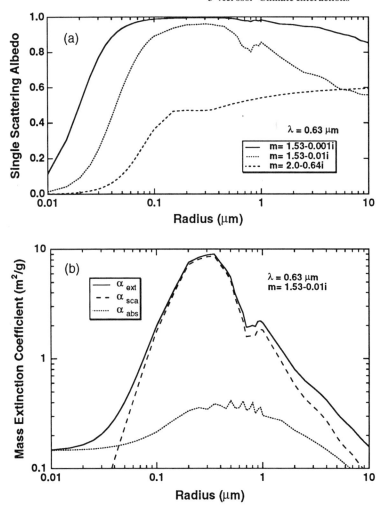

Figure 2   (a) Single scattering albedo of monodispersed spherical aerosols of varying radius and three different refractive indices at a wavelength of 0.63 μm. (b) Mass extinction (α_{ext}), scattering (α_{sca}), and absorption (α_{abs}) coefficients (in $m^2\ g^{-1}$) for the aerosol in (a) with refractive index $m = 1.53 - 0.01i$ and a particle density of 1 g cm^{-3}.

for single particles of varying radius. The three separate curves are for aerosols composed of carbon ($m = 2.0 - 0.64i$) and two models of sulfate aerosols containing absorptive components. Figure 2b shows the specific mass extinction coefficients, α_{ext}, α_{sca} and α_{abs}, at 0.63 μm for the aerosol of complex index $m = 1.53 - 0.01i$ assuming a monodispersed distribution of aerosols of given radius composed of material having a density of 1 g cm^{-3}.

The shortwave forcing of the clear sky radiation due to tropospheric aerosols can be estimated from Eq. (5), which is the perturbation in the albedo of the surface–aerosol system. To translate this into a change in planetary albedo, the effect of spectrally varying molecular scattering, stratospheric ozone absorption, and tropospheric water-vapor absorption needs to be considered. However, first-order effects can be obtained by assuming that the atmosphere is a grey nonreflecting layer overlying the surface–aerosol system (Charlson et al., 1991).

The perturbation over cloudy areas will be exceedingly small considering the disparity in optical depths. Figure 3 shows a zonal mean global climatology of aerosol visible optical depths obtained from several years of land-based transmission measurements (Weller and Leiterer, 1988). For comparison, monthly mean cloud optical depths from the International Satellite Cloud Climatology Project (ISCCP) C2 data (Rossow and Schiffer, 1991) are shown on the same panel.

The change in planetary albedo ΔR_p for a nonabsorbing aerosol layer can therefore be written by combining Eqs. (1) and (5) as

$$\Delta R_p \approx T_{atm}^2 (1 - N_c)(1 - R_s)^2 (2\beta\tau) \tag{7}$$

where T_{atm} is the transmittance of the grey atmosphere above the aerosol layer, N_c is the global mean cloud fraction, and other quantities are as previously defined. The annual global mean insolation of the earth–atmosphere system is $S_o/4$ where S_o is the solar constant, which is 1370 W m^{-2}. The planetary radiative forcing is then

$$\Delta F_R = \Delta R_p S_o /4 \tag{8}$$

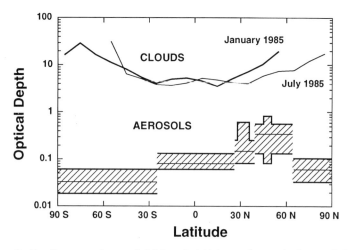

Figure 3 Zonally averaged mean of visible optical thickness of atmospheric aerosols from land-based transmission measurements (Weller and Leiterer, 1988) and mean cloud optical thickness for January and July 1985 from ISCCP C2 data.

For generally accepted values (Charlson et al., 1991) of $T_{atm} = 0.71$, $N_c = 0.6$, $R_s = 0.15$, and $\beta = 0.3$, Eq. (8) yields

$$\Delta F_R \approx 30\tau \tag{9}$$

which, for τ ranging from 0.05–0.10 (see Fig. 3), implies a radiative forcing of 1.5–3.0 W m^{-2}, a value comparable to the longwave radiative forcing of several minor greenhouse gases (Shine et al., 1990), but of opposite sign.

III. Early Applications

The role of aerosols in modifying the earth's climate through the alteration of the various components of the heat budget has been a topic of scientific discussion for many decades. There was considerable concern that an increase in atmospheric turbidity would lead to severe cooling of the earth's surface (SMIC, 1971). National and international turbidity networks were installed in the 1960s (Flowers et al., 1969) and there was extensive cooperation between American and Soviet scientists studying various aspects of aerosol–climate interactions (Kondratyev, 1991).

The importance given to aerosol forcing may be appreciated by realizing that climate research of that period was dominated by two events. The first was the eruption of Mt. Agung in 1963, which led to a dramatic increase in the stratospheric aerosol burden. Several studies related volcanic activity to subsequent temperature declines (Lamb, 1970); indeed, the northern hemisphere temperature showed a decline for the two years following the eruption of Mt. Agung. The second factor was the influence of the pioneering climate model studies of Budyko (1969) and Sellers (1969). Results from the application of the original Budyko–Sellers models showed that a reduction in solar input of as little as 2% could trigger an ice age. Although studies dealt primarily with stratospheric aerosols of volcanic origin, emphasis began shifting toward tropospheric aerosols, and in particular, aerosols of anthropogenic origin.

A. Background Aerosols

It was general practice to consider both natural and anthropogenic components of the aerosol burden together, and attempts were made at generating an aerosol climatology of this background layer based on observations. The model of Toon and Pollack (1976) was representative of conditions in the northern hemisphere and used extensively in model calculations of the effect of aerosols on climate. Although several studies were made, basically two types of simple climate models were used: the radiative–convective model, which resolves radiative perturbations in an atmospheric column, and the energy balance model, which allows for latitudinal dependence but parameterizes all processes in terms of the surface temperature.

The conclusions reached by Charlock and Sellers (1980) are typical for radiative–convective models. They used an enhanced one-dimensional model that includes the effects of meridional heat transport and heat storage. The model was run over the annual cycle for conditions at 40° and 50°N latitude without aerosols and with a prescribed aerosol optical depth of 0.125 at 0.55 μm. The annual mean surface temperature with aerosols was 1.6 K lower than that for the aerosol-free run. The model was also run in an annual mode without heat storage and showed a 1.8 K drop for the same conditions. When the model was run for global mean conditions, the temperature drop was 1.5 K, but inclusion of ice-albedo feedback through a temperature-dependent global albedo would have doubled the drop in temperature.

Coakley et al. (1983) computed the latitudinally dependent radiative forcing for the Toon and Pollack (1976) aerosol distribution, including the effects of absorbing components prior to estimating aerosol effects, using an energy balance model for northern hemisphere conditions. Even for moderately absorbing aerosols ($m = 1.5 - 0.01i$), the solar radiative forcing was negative except in the 80°–90°N latitude belt. The model results showed global mean surface temperature decreases ranging from 3.3 K for nonabsorbing aerosols to 2.0 K for the absorbing aerosol model with $k = 0.01$. Significantly, the maximum temperature drop was at polar latitudes, even for the absorbing layer that produced the positive radiative forcing. This is because high-latitude climatic change is controlled through advective processes that responded to the aerosol-induced cooling at low and mid latitudes. Other two-dimensional model studies confirmed this basic picture (Jung and Bach, 1987).

In a later study (Coakley and Cess, 1985), the background aerosol model was introduced into a general circulation model. However, since the model had fixed sea-surface temperature, the global mean surface temperature perturbation was negligible in spite of the fact that the negative radiative forcing at the top of the atmosphere was 5.0 W m^{-2} over the oceans and even higher over land.

B. Chronic Regional and Seasonal Effects

The study of aerosol effects at the local or regional scale also has a rich history, particularly with respect to precipitation. However, the radiative effect that attracted the most attention concerned Arctic aerosols and the accompanying haze layer (Rosen et al., 1981). A field experiment, the Arctic Gas and Aerosol Sampling Program (AGASP), was conducted in the spring of 1983 (Schnell, 1984). The aerosols were found to be composed of an absorbing component and hygroscopic material that led to increased haziness at high humidities.

The study by MacCracken et al. (1986), using both one- and two-dimensional climate models, summarizes the consequences of the radiative forcing by Arctic aerosols. The initial forcing of the surface–atmosphere system is positive for surface albedos higher than 0.17, and the equilibrium response of the one-dimensional radiative–convective model correlates well with the forcing showing surface temperature increases up to 8 K. They pointed out the role of infrared emission from the warmer atmosphere in forcing the surface since the direct solar forcing for the surface is actually negative. The two-dimensional model was run through the seasonal cycle and had interactive cryospheric processes. The increase in mean surface temperature in the 65°–90°N belt relative to the control peaked in May, a month later than the peak in radiative forcing as a result of earlier snow melt.

Another region subject to chronic moderate aerosol loads is the Saharan region where windblown dust is a seasonal feature (Carlson and Prospero, 1972). Carlson and Benjamin (1980) calculated the shortwave and longwave heating rates for a fairly absorbing thick dust layer and found an excess of 1 K day^{-1} in the 500–1000 mb net heating rate over dust-free conditions for optical depths of 1.0. This heating when inserted into a general circulation model suggested a narrowing and intensification of the Intertropical Convergence Zone and a weakening of the trade winds (Randall et al., 1984).

C. Episodic Massive Aerosol Loads

The study of the consequences of massive injections of aerosols has been dominated by simulations of the climatic effects of nuclear war. There is now nearly incontrovertible evidence that massive injections of debris into the earth's atmosphere have occurred in the past and that these episodes have coincided with the extinction of species (Alvarez et al., 1980; Claeys et al., 1992). The possibility that such a catastrophe could occur through means other than acts of God, namely nuclear war, was raised by Crutzen and Birks (1982). The climatic consequences of widespread nuclear war was further quantified by Turco et al. (1983). The measuring efforts that followed in pinning down the degree of the calamity that would result from large-scale nuclear war provided a better understanding of the nature and magnitude of smoke generated by various types of fires and the processes that cause the lofting, dispersion, and scavenging of smoke (e.g., Radke et al., 1988, 1990, 1991).

Modeling efforts related to what has become known as "nuclear winter" ranged from radiative–convective models (Cess et al., 1985) to three-dimensional general circulation models (Thompson et al., 1987; Ghan et al., 1988) and mesoscale models (Giorgi and Visconti, 1989) with interactive smoke generation and removal processes and fairly detailed smoke optics. An important point raised by Cess et al. (1985) was that the solar absorption in the smoke layer decouples the surface and atmosphere. A compendium of the atmospheric effects of nuclear war is contained in Pittock et al. (1986). A review of modeling efforts has been made by Schneider and Thompson (1988) and more recently by Turco et al. (1990). A summary of results from the latter indicates that for a July smoke injection, land temperatures averaged over the $30°-70°N$ latitude zones would decrease by 5 K, 13 K, and 22 K, respectively, for smoke absorption optical depths of 0.3, 1.0, and 3.0, during the initial acute phase of a postnuclear war catastrophe. Depending on the destruction scenarios, the estimated absorption optical depth is expected to be between 0.24 and 10.4 with a most likely value of 2.3.

IV. Recent Integrated Studies

It has now been realized that tropospheric aerosols should be included in atmospheric models as an additional radiatively active constituent. The most recent versions of the radiation parameterization in European general circulation models (GCMs) allows for the inclusion of prescribed aerosol layers (Morcrette, 1991; Ritter and Geleyn, 1992). Aerosol optical properties and distribution over geographical regions is based on consensus models (World Climate Programme, 1983; Tanre et al., 1984) that probably need to be reconsidered. The inclusion of aerosol effects may be justified on the grounds that the direct radiative forcing is comparable to that of minor trace gases that are now routinely included in GCMs. For instance, the latest version of the National Center for Atmospheric Research (NCAR) Community Climate Model incorporates the radiative effects of methane, oxides of nitrogen, and CFCs (Briegleb, 1992).

A. Natural Aerosols

The sources and sinks of tropospheric aerosols of natural origin are discussed in Chapter 2. Even in the absence of anthropogenic sources, aerosols have played a role in the atmospheric radiative budget. Recent studies have dealt mainly with the last glacial maximum (LGM) of about 18,000 years ago and the Cretaceous–Tertiary boundary (K/T). Harvey (1988) used an energy balance model to show that increased aerosol abundances during the LGM, as found in ice cores (Royer et al., 1983), could have contributed an additional 2–3 K cooling. However, Anderson and Charlson (1990) pointed out that the extra burden is mainly coarse-mode dust and sea salt that are not as optically efficient as fine-mode sulfate (see Fig. 2b); they felt that Harvey (1988) overestimated the aerosol effect by a factor of 2–10.

Anders et al. (1991) cite evidence of major wildfires at the K/T boundary. The size distribution is also compatible with sorting of smoke and ejecta during transit with some sites apparently far from the K/T impact site. Their studies are relevant to questions regarding the effect of biomass burning on the current tropospheric aerosol load and atmospheric radiative budget.

Aside from these anomalous conditions, the natural aerosol load of sulfate aerosol that provides the primary direct radiative forcing is much smaller than the anthropogenic burden, especially in the northern hemisphere. Figure 4 from

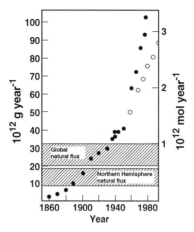

Figure 4 Time history of total global anthropogenic emissions of SO₂ (shown as grams and moles of sulfur), and estimates of total global and northern-hemisphere natural emissions of gaseous reduced-sulfur compounds and SO₂. (From Charlson et al., 1992; Copyright © 1992 by the AAAS.)

Charlson et al. (1992) shows the time history of relative emission rates of sulfur. Lately, studies have focused on estimating the direct radiative forcing of anthropogenic aerosols.

B.   Industrial Emissions

A large part of the submicrometer-sized tropospheric aerosol load is from gas-to-particle conversion through chemical processes involving sulfur and hydrocarbons. The primary anthropogenic source of sulfur emissions is fossil fuel combustion. Watson et al. (1990) estimate the global flux from this source to be 80 Tg S yr⁻¹ (1 Tg S = 10¹² g of sulfur). An additional 7 Tg S yr⁻¹ is from biomass burning.

Charlson et al. (1991) have computed the sulfate aerosol burden using a three-dimensional transport-chemistry model. Their results for the combined emissions from natural and anthropogenic sources are shown in Fig. 5. The aerosol column amount may be converted to optical depth using the appropriate mass extinction coefficient α_{ext}, which ranges from 5 to 10 m² g⁻¹ depending on the humidity, a range compatible with Fig. 2b. For a value of α_{ext} = 10 m² g⁻¹, the optical depth is numerically equal to one-hundredth the sulfate burden (in mg m⁻²). Figure 5 is then a global map of the aerosol optical depth and, in fact, compares quite favorably in the zonal mean with the measured values shown in Fig. 3. The global mean value of the sulfate burden for the anthropogenic component alone is also compatible with estimates based on sulfur emission, sulfate conversion, and tropospheric lifetimes (Charlson et al., 1992). The corresponding optical depth is 0.04, so that the direct radiative forcing is 1.2 W m⁻² from Eq. (9). This value is uncer-

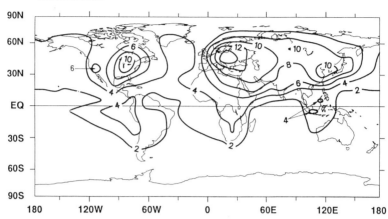

Figure 5   Modeled annual mean distributions of the column burden of sulfate aerosols including natural and anthropogenic sources. The contours (in mg m^{-2} of sulfate) can also be interpreted as the visible optical depth in hundredths. (From Charlson et al., 1991; Copyright © 1991 by Munskgaard International Publishers Ltd.)

tain by a factor of 2, but from the trend in Fig. 4 the forcing appears to be increasing by 0.2 W m^{-2} per decade. For reference, the corresponding trend for carbon dioxide is 0.3 W m^{-2} (Watson et al., 1990) but the increase is logarithmic in CO_2 concentration whereas the aerosol forcing is linear, as can be seen from Eq. (9).

Policies aimed at reducing sulfur emissions will result in an immediate decrease in the upward trend in radiative forcing due to sulfate aerosols. Curbing fossil fuel combustion will not immediately reduce the increase in atmospheric carbon dioxide content, because of the long lifetime of the carbon cycle. The implication of this led Wigley (1989, 1991) to construct possible future scenarios in which the sulfate forcing was reduced. He had detected a slight interhemispheric difference in the temperature record over the past century that he attributed to greater indirect aerosol–cloud radiative forcing in the northern hemisphere. Earlier, Schwartz (1988, 1989) had been unable to find any such signal in cloud albedo or temperature data. Interhemispheric differences in cloud cover and meteorology make such comparisons somewhat questionable in any case. Nevertheless, the cancellation in sulfate and greenhouse gas forcing is asymmetric, and Wigley (1989) pointed out that the climate response to asymmetric shortwave forcing may be quite different from the response to the global greenhouse forcing. General circulation models respond similarly to equivalent shortwave and longwave forcing when the forcing is uniform, such as for a change in the solar constant or an increase in atmospheric carbon dioxide concentration (Manabe and Wetherald, 1980), but these models have not been used to study asymmetric, combined shortwave and longwave forcing. In any case, changes in cloud cover could completely swamp the sulfate aerosol signal.

C. Biomass Burning

Biomass burning has been a feature of the environment since the evolution of land plants. Currently, natural causes are being supplanted by deliberate human action on an unprecedented scale (Andreae, 1991). The pollutants emitted in the process result in the formation of tropospheric aerosols that add significantly to the radiative forcing attributed to sulfates. Apart from sulfur gases that add to the sulfate aerosol burden, smoke from biomass burning consists of carbonaceous particles that are both scattering and absorbing (Mazurek et al., 1991). The radiative forcing may be determined by estimating the global burden of smoke aerosol and its radiative properties.

Penner et al. (1991, 1992) have gone through the various steps needed to arrive at an estimate of the global radiative forcing. The total mass of carbon burned annually has been estimated at 3500 Tg (Crutzen and Andreae, 1990; Andreae, 1991). The smoke produced by the burning is determined by the particle emission factor, which has been measured to be $20-35$ g kg^{-1} (Radke et al., 1991). The best estimate of the total mass of smoke produced is therefore about 90 Tg yr^{-1}, which is comparable to the anthropogenic sulfate production shown in Fig. 4. For a tropospheric lifetime of 6 days, which is typical of sulfate aerosols of similar size (Charlson et al., 1991), the smoke loading is 3.6×10^{-3} g m^{-2}.

The radiative properties of the layer are determined by its optical properties. Radke et al. (1988, 1991) found that the single scattering albedo ranges from a low value of 0.7 following ignition to about 0.9 for an established fire, with the average for several cases of 0.83 ± 0.11. The specific mass absorption coefficient α_{abs} was 0.64 ± 0.36 m^2 g^{-1}. Golitsyn et al. (1988) cite a value of $\varpi = 0.72$ and $\alpha_{ext} = 5.6$ m^2 g^{-1} from measurements for smoke. The single scattering albedo depends on the ratio of organic carbon to elemental carbon, which varies with the intensity and duration of the fire (Mazurek et al., 1991).

A synthesis of all of the above information allowed Penner et al. (1992) to estimate that the global mean extinction optical depth due to smoke aerosols is 0.03, which is comparable to the estimate of sulfate aerosol optical depth made by Charlson et al. (1992). For purely scattering aerosols, the radiative forcing from Eq. (9) is then 0.9 W m^{-2}. Smoke absorption will result in a positive atmospheric forcing, which was estimated by Penner et al. (1992) to be 0.5 W m^{-2}, with the net surface–atmosphere forcing being 0.2 W m^{-2}.

The combined shortwave radiative forcing due to sulfate aerosols and smoke is then more than 2 W m^{-2}, which is of roughly the same magnitude as the longwave forcing of the increase in concentration of all greenhouse gases since the pre-industrial era. This suggests that the effect on climate of fossil fuel and biomass burning can be ascertained only by taking an integrated approach (Kaufman et al., 1991).

D. Regional Catastrophes

Previous sections have dealt with the global radiative forcing caused by the combined effect of distributed sources of aerosols. Occasionally, natural or anthropogenic events create locally severe aerosol loads that could lead to large climatic perturbations. One example is the springtime Arctic haze that has been studied for several years. Recent work has been primarily directed toward forcing climate models with modeled radiative perturbations. Blanchet (1989, 1991) introduced increasingly heavy aerosol loads north of 60°N into a GCM to study the climatic response at higher latitudes. Although the solar heating rate in the troposphere increased quite dramatically, the temperature did not rise substantially for the nominal to 10 times nominal runs. The positive forcing of 0.1–0.3 K day^{-1} results in a decrease in the meridional heat flux in line with the findings of earlier simple models. More important, the cloud cover in the experiment was altered sufficiently that the changes in net radiative fluxes at the top were locally an order of magnitude greater than the initial forcing. This has profound implication for any observational or modeling study attempting to detect the aerosol signal. Another high-latitude phenomenon related to aerosols is the alteration of surface albedo due to the deposition of soot; this has been studied by Vogelmann et al. (1988) in connection with the nuclear winter hypothesis.

Several studies have examined the systematic effect of smoke from forest fires on air temperature near the surface. Veltishchev et al. (1988) analyzed data from several meteorological stations to study the effect of major historical fires in Siberia, Europe, and Canada. They estimated that the optical depth of smoke following fires in Siberia in 1915 was about 3.0 with surface temperature drops of 5 K in station data. Robock (1988) examined the situation in northern California where a subsidence inversion trapped smoke in mountain valleys for several days in September 1987. One station recorded a maximum temperature anomaly of −20°C.

Other studies have examined the effect of more widespread forest-fire smoke. Ferrare et al. (1990), Robock (1991), and Westphal and Toon (1991) studied smoke from Canadian fires in July 1982. The extensive smoke plume was visible in satellite imagery for several days. Robock (1991) estimated the effect of the smoke layer by comparing forecasted temperatures with the observations. He found that regions of negative anomaly were well correlated with the smoke layer. Westphal and Toon (1991) simulated the effects of the fire with a mesoscale model that included interactive smoke physics and optics. Their simulated optical depths matched the satellite-derived estimate of as high as 1.8 obtained by Kaufman et al. (1990). They calculated the albedo of the smoke-covered area to be 35%, and the resulting surface cooling was 5 K.

An opportunity to study the effect of heavy aerosol burdens arose following military action in Kuwait in 1991 when the withdrawing Iraqi army set fire to oil wells, petroleum storage tanks, and refineries. In answer to concerns that the resulting smoke plume would cause global climatic perturbations, several modeling studies were undertaken (Small 1991; Browning et al., 1991; Bakan et al., 1991). The basic issue was whether smoke would reach the stratosphere and cover a significant geographical area. Studies of smoke from oil burning carried out earlier (Radke et al., 1990) had shown the possibility of self-lofting of smoke as a result of solar heating.

Browning et al. (1991) simulated the smoke plume with a long-range dispersion model and concluded that the smoke would remain in the troposphere. Beneath the plume, within about 200 km of the source, the maximum temperatures were expected to drop by about 10 K. Bakan et al. (1991) used a GCM with an interactive tracer model to simulate the plume dispersion and climatic effects. They computed fields of soot deposition and surface temperature change. On a regional scale, maximum temperature drops were estimated to be about 4 K near the source.

Airborne studies of the smoke from the Kuwait fires were also carried out (Johnson et al., 1991; Hobbs and Radke, 1992). The smoke was observed to remain below 5–6 km and to be confined primarily to the Gulf region [although there were some unsubstantiated claims of transport over much longer distances (Bodhaine et al., 1992; Deshler and Hofmann, 1992)]. The optical depth near the source was 2 to 3 and the single scattering albedo ranged from 0.35 for the blackest plumes to 0.95 from white plumes that contained a fair amount of salt (Hobbs and Radke, 1992). Overall the smoke contained less soot and had a higher single scattering albedo than expected. Also, many of the smoke particles were cloud condensation nuclei, so that their residence time in the atmosphere would have been only a few days. These properties of the smoke particles, together with their limited transport, resulted in the smoke from the Kuwait fires having negligible effects globally.[1]

The study of catastrophic events such as those described above is useful for model verification, since such a strong forcing usually leads to an unambiguous response. By contrast, the response of models to the radiative forcing of the global distributions of natural and (normal) anthropogenic sulfate or smoke is well within the limits of internal model variability; therefore, a clear signal may be difficult to discern.

V. Outstanding Problems

The gaps in our understanding of the tropospheric aerosol–climate response problem are in both modeling and monitoring. Developments on both fronts should proceed in parallel.

---

[1] The reader is referred to the *Journal of Geophysical Research* (1992, **97**, 14,481–14,580) for a comprehensive set of papers on the smoke of the Kuwait oil fires.

A. Modeling

With reference to Fig. 1, the uncertainties that lie within the shaded box representing the direct radiative forcing due to aerosols are such that the forcing of the earth–troposphere system is known to within a factor of 2 or 3. The estimates provided in this review indicate that even at the low end the forcing warrants inclusion in GCMs if the total anthropogenic radiative forcing of the earth's climate is to be considered. The chemical pathways by which sulfates and smoke are produced is fairly well understood and supported by extensive observations. The same cannot be said of the indirect forcing through changes in cloud microphysics.

The relationship between aerosol mass load and radiative properties is also quite robust. The one recent development that deserves mention is the application of the theory of the optics of fractal clusters (Berry and Percival, 1986) to the problem of computing extinction coefficients for sooty smoke particles that occur as chained aggregates of small spherules. Nelson (1989) has shown that the mass scattering and especially mass absorption coefficients of these particles do not decrease with increasing size, as is the case for spherical particles. (Note the decay with increasing size shown in Fig. 2b.) Instead, the aggregate retains the character of the individual small spherules. Therefore, coagulation with aging does not diminish the extinction properties of a smoke layer, although it will change the atmospheric residence time of the particles. Additional direct radiative effects of tropospheric aerosols should be mentioned—for instance, the scattering of biologically active ultraviolet radiation (Liu et al., 1991). This is probably more important locally than on a global scale.

The primary hurdle toward a better understanding of the problem lies in the modeling of climate feedbacks, especially through changes in cloudiness. Figure 3 serves as a stark reminder that the cloud radiative forcing can overwhelm any aerosol effects. The current global average cloud shortwave radiative forcing is 50 W m^{-2} (Harrison et al., 1990) and a 1% relative change in the cloud cover corresponds to a forcing of 0.5 W m^{-2}. The range in cloud shortwave radiative forcing simulated by current GCMs exceeds the total aerosol forcing by an order of magnitude (Harrison et al., 1990), which suggests that models will be hard pressed to yield a clear signal in response to the aerosol forcing.

B. Process Studies

Modeling of the radiative effects of aerosols requires an accurate estimate of the global aerosol burden and its relationship to industrial and other human activities. To accomplish this, uncertainties in the particle emission rates, aerosol residence times, and dispersion, deposition, and coagulation rates need to be reduced. The mesoscale model studies (Giorgi and Visconti, 1989; Westphal and Toon, 1991) and GCM simulations (Browning et al., 1991; Bakan et al., 1991) cited in this review incorporate these processes through parameterizations of varying complexity. The International Global Aerosol Program (IGAP), just getting underway, is expected to address these issues.

C. Monitoring

The relentless rise in greenhouse gas concentrations has been monitored globally since the mid-1950s. There has been no corresponding continuous program for aerosols. The highly variable distribution requires a truly global effort, and space-based observing systems are the hope of the future. Currently, NOAA/NESDIS prepares weekly global maps of the aerosol optical depth retrieved from AVHRR channel 1 (about 0.58–0.68 μm) upwelling radiances over the ocean (Rao et al., 1989; Stowe, 1991). An example is shown in Fig. 6 for the week of May 23–29, 1991. The enhanced load in the tropics is apparently genuine and is primarily windblown dust. It is not a result of cloud contamination, which is removed using information from the thermal channels (McClain, 1989). The most striking feature is the stark interhemispheric asymmetry in the aerosol optical depth. The World Meteorological Organization, under its Global Atmospheric Watch Program, is currently engaged in expanding the network of ground-based aerosol-measuring stations. Data from this effort will complement the remote sensing program.

Aerosol monitoring is also a vital function of the MODIS-N instrument on EOS (Earth Observing System). Algorithms are in place to extract optical depth, particle size, and single scattering albedo (King et al., 1992). Whereas current operational techniques are limited to oceanic areas, MODIS channels are expected to provide aerosol optical depths over vegetation and dark soils. The finer spatial resolution of the sensor is also expected to minimize cloud contamination. The methods have so far been used successfully for a few individual cases of obvious pollution (Kaufman et al., 1990).

An estimate of the global distribution of the visible optical depth obtained through the appropriate blending of remotely sensed data and surface-based observations is sufficient to estimate the shortwave radiative forcing to first order. The absorptive component of the aerosol burden is important locally in the Arctic but is quite small globally, and the reliability of retrieved or measured single scattering albedos is probably not crucial at this stage of the analysis of aerosol–climate effects. The partition of the aerosol burden between natural and anthropogenic sources could be accomplished through the analysis of regional trends. The stratospheric component, which is primarily of volcanic origin, is already being mapped using techniques discussed in Chapter 8 and can be subtracted out. As in the case of modeling, the success of any attempt at monitoring trends in aerosol burden will ultimately rest on our ability to isolate the aerosol signal from the much more dominant effect of changes in cloudiness.

In summary, there is hope that modeling and monitoring efforts will converge to provide a clearer picture of the contribution of aerosols to the earth's radiation

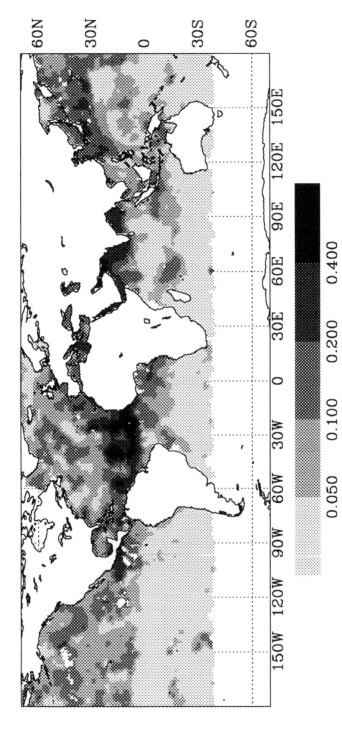

Figure 6 The mean visible optical depth of aerosols over oceanic areas for the week of May 23–29, 1991 retrieved by NOAA/NESDIS from AVHRR channel 1.

budget. Well-established programs for studying the role of clouds in climate forcing will also help in isolating aerosol effects. In any case, following the discussion in Chapter 2 of this volume, aerosols and clouds need to be considered together as a system. Finally, aerosol radiative forcing is proof, if any was needed, that human activity is inextricably linked with the future of the earth's climate.

Acknowledgments

Support through NSF Grants ATM-8909870 and ATM-9002936 and NASA Grant NAG5-1125 is gratefully acknowledged. I wish to thank Wanda Curtis and Tim Gilbert for assistance with the preparation of this manuscript, Ray Espinoza for the scattering computations, and Grant Petty for the map of aerosol optical depth, which was plotted from data kindly provided by Larry Stowe of NOAA/NESDIS.

References

Ackerman, T. P., and Toon, O. B. (1981). *Appl. Opt.* **20**, 3661–3667.
Alvarez, L. W., Alvarez, W., Asaro, F., and Michel, H. V. (1980). *Science* **208**, 1095–1108.
Anders, E., Wolbach, W. S., and Gilmour, I. (1991). In *Global Biomass Burning: Atmospheric, Climatic, and Biospheric Implications* (J. S. Levine, ed.), 485–492. The MIT Press, Cambridge, Massachusetts.
Anderson, T. L., and Charlson, R. J. (1990). *Nature* **345**, 393.
Andreae, M. O. (1991). In *Global Biomass Burning: Atmospheric, Climatic, and Biospheric Implications* (J. S. Levine, ed.), 3–21. The MIT Press, Cambridge, Massachusetts.
Bakan, S., Chlond, A., Cubasch, V., Feichter, J., Graf, H., Grassl, H., Hasselmann, K., Kirchner, I., Latif, M., Roeckner, E., Sausen, R., Schlese, U., Schriever, D., Schult, I., Schumann, U., Sielman, F., and Welke, W. (1991). *Nature* **351**, 367–371.
Berry, M. V., and Percival, I. C. (1986). *Optica Acta* **33**, 577–589.
Blanchet, J.-P. (1989). *Atmos. Environ.* **23**, 2609–2625.
Blanchet, J.-P. (1991). In *Pollution of the Arctic Atmosphere* (W.T. Sturges, ed.), 289–322. Elsevier.
Bodhaine, B. A., Harris, J. M., Ogren, J. A., and Hofmann, D. J. (1992). *Geophys. Res. Lett.* **19**, 581–584.
Bohren, C. F., and Huffman, D. R. (1983). *Absorption and Scattering of Light by Small Particles.* John Wiley, New York, 530 pp.
Briegleb, B. P. (1992). *J. Geophys. Res.* **97**, 11,475–11,485.
Browning, K. A., Allam, R. J., Ballard, S. P., Barnes, R. T. H., Bennetts, D. A., Maryon, R. H., Mason, P. J., McKenna, D., Mitchell, J. F. B., Senior, C. A., Slingo, A., and Smith, F. B. (1991). *Nature* **351**, 363–367.
Budyko, M. I. (1969). *Tellus* **21**, 611–619.
Carlson, T. N., and Prospero, J. M. (1972). *J. Appl. Meteor.* **11**, 283–297.
Carlson, T. N., and Benjamin, S. G. (1980). *J. Atmos. Sci.* **37**, 193–213.
Cess, R. D., Potter, G. L., Ghan, S. J., and Gates, W. L., (1985). *J. Geophys. Res.* **90**, 12,937–12,950.
Charlock, T. P., and Sellers, W. D. (1980). *J. Atmos. Sci.* **37**, 1327–1341.
Charlson, R. J., Langner, J., Rodhe, H., Leovy, C. B., and Warren, S. G. (1991). *Tellus* **43AB**, 152–163.
Charlson, R. J., Schwartz, S. E., Hales, J. M., Cess, R. D., Coakley, J. A., Jr., Hansen, J. E., and Hofmann, D. J. (1992). *Science* **255**, 423–430.

Chylek, P., and Coakley, J. A., Jr. (1974). *Science* **183**, 75–77.

Claeys, P., Casier, J.-G., and Margolis, S. (1992). *Science* **257**, 1102–1104.

Coakley, J. A., Jr., and Chylek, P. (1975). *J. Atmos. Sci.* **32**, 409–418.

Coakley, J. A., Jr., Cess, R. D., and Yurevich, F. B. (1983). *J. Atmos. Sci.* **40**, 116–138.

Coakley, J. A., Jr., and Cess, R. D. (1985). *J. Atmos. Sci.* **42**, 1677–1692.

Crutzen, P. J., and Birks, J. W. (1982). *Ambio* **11**, 114–125.

Crutzen, P. J., and Andreae, M. O. (1990). *Science* **250**, 1669–1678.

Deshler, T., and Hofmann, D.J. (1992). *Geophys. Res. Lett.* **19**, 385–388.

Ferrare, R. A., Fraser, R. S., and Kaufman, Y. J. (1990). *J. Geophys. Res.* **95**, 9911–9925.

Flowers, E. C., McCormick, R. A., and Kurfis, K. R. (1969). *J. Appl. Meteor.* **8**, 955–962.

Ghan, S. J., MacCracken, M. C., and Walton, J. J. (1988). *J. Geophys. Res.* **93**, 8315–8337.

Giorgi, F., and Visconti, G. (1989). *J. Geophys. Res.* **94**, 1145–1163.

Golitsyn, G. S., Shukurov, A. K., Ginzburg, A. S., Sutugin, A. G., and Andronova, A. V. (1988). *Izv. Acad. Sci. USSR Atmos. Oceanic Phys.* **24**, 163–168.

Hansen, J. E., and Lacis, (1990). *Nature* **346**, 713–719.

Harrison E. F., Minnis, P., Barkstrom, B. R., Ramanathan, V., Cess, R. D., and Gibson, G. G. (1990). *J. Geophys. Res.* **95**, 18,687–18,703.

Harvey, L. D. (1988). *Nature* **334**, 333–335.

Hobbs, P. V., and Radke, L. F. (1992). *Science* **256**, 987–991.

Johnson, D. W., Kilsby, C. G., McKenna, D. S., Saunders, R. W., Jenkins, G. J., Smith, F. B., and Foot, J. S. (1991). *Nature* **353**, 617–621.

Jung, H. J., and Bach, W. (1987). *Theor. Appl. Climatol.* **38**, 222–233.

Kaufman, Y. J., Fraser, R. S., and Ferrare, R. A. (1990). *J. Geophys. Res.* **95**, 9895–9909.

Kaufman, Y. J., Fraser, R. S., and Mahoney, R. L. (1991). *J. Climate* **4**, 578–588.

King, M. D., and Harshvardhan (1986). *J. Atmos. Sci.* **43**, 784–801.

King, M. D., Kaufman, Y. J., Menzel, W. P., and Tanre, D. (1992). *IEEE Trans. Geosci. Remote Sensing* **30**, 2–27.

Kondratyev, K. Y. (1991). *Aerosols and Climate*. Leningrad Gidrometeoizdat, 541 pp.

Lamb, H. H. (1970). *Phil. Trans. Roy. Soc. London* **266**, 425–533.

Liu, S. C., McKeen, S. A., and Madronich, S. (1991). *Geophys. Res. Lett.* **18**, 2265–2268.

MacCracken, M. C., Cess, R. D., and Potter, G. L. (1986). *J. Geophys. Res.* **91**, 14,445–14,450.

Manabe, S., and Wetherald, R. T. (1980). *J. Atmos. Sci.* **37**, 99–118.

Mazurek, M. A., Cofer, W. R. III, and Levine, J. S. (1991). In *Global Biomass Burning: Atmospheric, Climatic, and Biospheric Implications* (J. S. Levine, ed.), 258–263. The MIT Press, Cambridge, Massachusetts.

McClain, E. P. (1989). *Int. J. Remote Sens.* **10**, 763–769.

Morcrette, J.-J. (1991). *J. Geophys. Res.* **96**, 9121–9132.

Nelson, J. (1989). *Nature* **339**, 611–613.

Penner, J. E., Ghan, S. J., and Walton, J. J. (1991). In *Global Biomass Burning: Atmospheric, Climatic, and Biospheric Implications* (J. S. Levine, ed.), 387–393. The MIT Press, Cambridge, Massachusetts.

Penner, J. E., Dickinson, R. E., and O'Neill, C. A. (1992). *Science* **256**, 1432–1433.

Pittock, A. B., Ackerman, T. P., Crutzen, P. J., MacCracken, M. C., Shapiro, C. S., and Turco, R. P. (1986). *Environmental Consequences of Nuclear War, Vol. I: Physical and Atmospheric Effects*. SCOPE Publication No. 28. John Wiley, New York.

Radke, L. F., Hegg, D. A., Lyons, J. H., Brock, C. A., Hobbs, P. V., Weiss, R. E., and Rasmussen, R. (1988). In *Aerosols and Climate* (P. V. Hobbs and M. P. McCormick, eds.), 411–422. A. Deepak Publishing, Hampton, Virginia.

Radke, L. F., Lyons, J. H., Hobbs, P. V., and Weiss, R. E. (1990). *J. Geophys. Res.* **95**, 14,071–14,076.

Radke, L. F., Hegg, D. A., Hobbs, P. V., Nance, J. D., Lyons, J. H., Laursen, K. K., Weiss, R. E.,

Riggan, P. J., and Ward, D. E. (1991). In *Global Biomass Burning: Atmospheric, Climatic, and Biospheric Implications* (J. S. Levine, ed.), 209–224. The MIT Press, Cambridge, Massachusetts.

Randall, D. A., Carlson, T. N., and Mintz, Y. (1984). In *Aerosols and their Climatic Effects* (H. E. Gerber and A. Deepak, eds.), 123–132. A. Deepak Publishing, Hampton, Virginia.

Rao, C. R. N., Stowe, L. L., and McClain, E. P. (1989). *Int. J. Remote Sens.* **10**, 743–749.

Ritter, B., and Geleyn, J.-F. (1992). *Mon. Wea. Rev.* **120**, 303–325.

Robock, A. (1988). *Science* **242**, 911–913.

Robock, A., (1991). *J. Geophys. Res.* **96**, 20,869–20,878.

Rosen, H., Novakov, T., and Bodhaine, B. A. (1981). *Atmos. Environ.* **15**, 1371–1375.

Rossow, W. B., and Schiffer, R. A. (1991). *Bull. Am. Met. Soc.* **72**, 2–20.

Royer, A., DeAngelis, M., and Petit, J. R. (1983). *Climatic Change* **5**, 381–412.

Schneider, S. H., and Thompson, S. L. (1988). *Nature* **333**, 221–227.

Schnell, R. C. (1984). *Geophys. Res. Lett.* **11**, 359.

Schwartz, S. E. (1988). *Nature* **336**, 441–445.

Schwartz, S. E. (1989). *Nature* **340**, 515–516.

Sellers, W. D. (1969). *J. Appl. Meteorol.* **8**, 392–400.

Shine, K. P., Derwent, R. G., Wuebbles, D. J., and Morcrette, J.-J. (1990). In *Climate Change: The IPCC Scientific Assessment* (J. T. Houghton, G. J. Jenkins, and J. J. Ephraums, eds.), 41–68. Cambridge University Press, Cambridge.

Small, R. D. (1991). *Nature* **350**, 11–12.

SMIC (1971). *Inadvertent Climate Modification: Report of the Study of Man's Impact on Climate (SMIC)*. The MIT Press, Cambridge, Massachusetts.

Stowe, L.L. (1991). *Palaeogeogr. Paleoclimatol. Palaeoecol.* **90**, 25–32.

Tanre, D., Geleyn, J.-F., and Slingo, J. (1984). In *Aerosols and their Climatic Effects* (H. E. Gerber and A. Deepak, eds.), 133–177. A. Deepak Publishing, Hampton, Virginia.

Thompson, S. L., Ramaswamy, V., and Covey, C. (1987). *J. Geophys. Res.* **92**, 10,942–10,960.

Toon, O. B., and Pollack, J. B. (1976). *J. Appl. Meteorol.* **15**, 225–246.

Turco, R. P., Toon, O. B., Ackerman, T. P., Pollack, J. B., and Sagan, C. (1983). *Science* **222**, 1283–1292.

Turco, R. P., Toon, O. B., Ackerman, T. P., Pollack, J. B., and Sagan, C. (1990). *Science* **247**, 166–176.

Veltishchev, N. N., Ginzburg, A. S., and Golitsyn, G. S. (1988). *Izv. Acad. Sci. USSR Atmos. Oceanic Phys.* **24**, 217–223.

Vogelmann, A. M., Robock, A., Ellingson, R. G. (1988). *J. Geophys. Res.* **93**, 5319–5332.

Watson, R. T., Rodhe, H., Oeschger, H., and Siegenthaler, V. (1990). In *Climate Change: The IPCC Scientific Assessment* (J. T. Houghton, G. J. Jenkins, and J. J. Ephraums, eds.), 1–40. Cambridge University Press, Cambridge.

Weller, M., and Leiterer, V. (1988). *Contrib. Atmos. Phys.* **61**, 1–9.

Westphal, D. L., and Toon, O. B. (1991). *J. Geophys. Res.* **96**, 22,379–22,400.

Wigley, T. M. L. (1989). *Nature* **339**, 365–367.

Wigley, T. M. L. (1991). *Nature* **349**, 503–506.

World Climate Programme (WCP), (1983). *Report of the Experts Meeting on Aerosols and Their Climatic Effects, Williamsburg, Virginia, WCP 55*. WMO-CAS and IAMAP Radiation Commission, Geneva, 107 pp.

Chapter 4 | Microphysical Structures of Stratiform and Cirrus Clouds

Andrew J. Heymsfield
National Center for Atmospheric Research
Boulder, Colorado

This chapter summarizes measurements of the physical and microphysical structures of stratus, stratocumulus, altostratus, and cirrus clouds, which are thought to have the greatest effect on climate because of the large areas they cover. These clouds can be loosely classified according to the phase(s) of the condensate: liquid, mixed-phase, and ice, respectively. Summaries of the microphysical properties, frequencies of occurrence, physical appearances, and heights and thicknesses of each of these cloud types are presented. The chapter draws on previously reported as well as unreported measurements.

I. Introduction

Clouds cover approximately 60% of the earth's surface; they strongly influence the global energy budget by partially controlling the amount of solar radiation absorbed by the earth and by partitioning this energy between the atmosphere and the earth's surface. Clouds also affect the loss of energy to space from thermal emissions from the earth. Ramanathan et al. (1983) and Ramanathan and Collins (1991) have proposed important feedback mechanisms between clouds and climate that involve the microphysical and radiative properties of clouds.

The radiative properties of a cloud depend primarily on its physical dimensions, its height in the atmosphere, and its microphysical characteristics. This chapter provides an overview of the microphysical structures of stratiform (stratus, stratocumulus, altostratus, and cirrus) clouds in the context of their potential effects on the earth's climate. For reviews of the mechanisms involved in droplet and ice particle growth in clouds, the reader is referred to Pruppacher and Klett (1978), Rogers and Yau (1989), Beard (1987), and Cooper (1991).

II.   Cloud Types and Occurrence

The three most common types of clouds are stratus, cumulus, and cirrus. According to the Glossary of Meteorology (Huschke, 1970), stratus clouds are layers or patches of low, often grey, clouds that have very little definition and rarely produce precipitation. Cumulus clouds are detached and dense, rising in mounds and towers from a level base. Cirrus clouds are high and of a silken appearance, a result of their ice crystal composition. These three primary designations are often combined; for instance, a low-level cloud that is broken up in a wave pattern is referred to as stratocumulus and a mid-level cloud displaying the same characteristics, as an altocumulus. A thick, mid-level layer cloud is termed an altostratus and a high-level cloud with a similar structure, cirrostratus. Precipitating clouds contain the word "nimbus." For example, a layer cloud producing precipitation is termed nimbostratus, and a deep precipitating cloud is called a cumulonimbus.

Clouds can also be classified by their structure into those consisting of liquid water (sometimes called "warm" clouds), those consisting of liquid water and ice ("mixed-phase" clouds), and those consisting of only ice ("ice" clouds). "Maritime" clouds and "continental" clouds are distinguished on the basis of geographical location and droplet concentration, the former being characterized by much lower droplet concentrations and larger droplets than the latter.

Table 1 shows the frequencies of the most commonly reported cloud types. The most extensive cloud cover over both land and sea consists of a combination of various types of stratus, altostratus, and cirrus. Therefore, this chapter will focus on the microphysical properties of these clouds, since they are likely to have the greatest effect on the radiation balance of the earth.

III.   Liquid Water Clouds

This section will present measurements in stratus (St) and stratocumulus (Sc) clouds. The data described in this and other sections were collected in several locations with the use of differing instrumentation. The reader should refer to the referenced articles for a description of the instrumentation employed.

A.   Stratus (St)

Observations of the microphysical structure of marine stratiform cloud layers (e.g., Noonkester, 1984) have shown many common features. Horizontally averaged liquid water contents increase with height less than adiabatically because of entrainment, which has the greatest effect near cloud top; the droplet concentration remains approximately constant throughout much of the cloud; and the mean size of the droplets increases monotonically with height. The physical dimensions of these clouds and their microphysical composition are summarized in

Table 1

Reports of the Most Common Cloud Types

(a) Oceanic Areas

Type	Frequency of occurrence (%)	Areal coverage over oceans (%)
Stratus (St) and Stratocumulus (Sc)	45	34
Cumulus (Cu)	33	12
Cumulonimbus (Cb)	10	6
Nimbostratus (Ns)	6	6
Altostratus (As) and Altocumulus (Ac)	46	22
Cirrus (Ci), Cirrostratus (Cs) and Cirrocumulus (Cc)	37	13
GLOBAL AVERAGE OVER OCEANS		64.8

Source: Warren et al., 1986.

Table 1

Reports of the Most Common Cloud Types

(b) Land Areas

Type	Frequency of occurrence (%)	Areal coverage over oceans (%)
St and Sc	27	18
Cu	14	5
Cb	7	4
Ns	6	5
As/Ac	35	21
Ci/Cs/Cc	47	23
GLOBAL AVERAGE OVER LAND		52.4

Source: Warren et al., 1988.

Table 2

Microphysical Structures of Stratus Clouds

Reference	Total droplet concentration (cm^{-3}) [a]	Liquid water content $(g\ m^{-3})$	Mean droplet volume diameter (μm)	Location/ comments
Herman & Curry (1984)	Up to 500	0.09 to 0.63	Up to 19	Arctic stratus. 6 flights.
Hobbs & Rangno (1985)	312	—	—	In and near Washington State. 4 cases.
Hudson (1983)	350 Marine 500 Continental	— —	— —	Off California coast. 1 Case. Droplet data begin above 0.46 μm diameter.
Nicholls & Leighton (1986)	~100–250	0.10 to 0.90	~12 near cld. base ~22 near cld. top	Over United Kingdom. 6 flights.
Noonkester (1984)	~200–350 Marine ~200–700 Continental	~0.3 maximum	~11 marine, ~7 continental	Off coast of California. 8 cases studied. Excellent resolution of the data in the vertical. Droplet data begin above 0.46 μm diameter.

[a] Above about 2 μm diameter.

Tables 2–6. The data for both St and Sc clouds are grouped together in some instances to increase the size of the data set. The designated cloud types are as given in the references. The data from Hobbs and Rangno (1985) are for maritime and continental clouds sampled over and in the vicinity of Washington State; this data set (for stratus at least) is small compared with those from the other locations (as reported in Borovikov et al., 1963).

Case-study analyses have unraveled some of the important physical processes operative in stratus clouds. Nicholls and Leighton (1986) investigated the microphysical, radiative, and turbulent properties of six stratiform clouds topping the marine boundary layers over the North Sea. They found the increase of liquid water with height to be within 10% of the adiabatic value except near cloud top,

Table 3
Physical Characteristics of Stratus and Stratocumulus Clouds

(a) Average altitude of cloud base (km)

Clouds	USSR		West Germany			England		Finland		India		
	Moscow	Leningrad	Hamburg	Köln	Karlsrühe	Mildenhall	Susterberg	Helsinki	Sodankyala	Calcutta	Madras	Simba
Sc	1.07	1.24	0.98	1.36	1.55	1.22	0.90	1.12	1.10	2.1	2.8	2.0
St	0.47	1.19	0.85	0.87	0.70	0.96	0.69	0.98	0.35	1.2	1.0	1.4

(b) Average thickness (km)

Clouds	Washington St.	Moscow	Hamburg	Köln	Mildenhall	Aldergrove	Susterberg	Lindenberg
Sc	1.23	0.36	0.46	0.38	0.44	0.52	0.40	0.31
St	0.38	0.52	0.46	0.36	0.74	0.54	0.50	0.32

Sources: Borovikov et al., 1963; Hobbs and Rangno, 1985.

Table 4

Liquid Water Content of Stratus and Stratocumulus
Clouds

Liquid water content ($g\ m^{-3}$)	St, Sc, (% of 326 observations)
0.00–0.09	12
0.10–0.19	32
0.20–0.29	22
0.30–0.39	16
0.40–0.49	12
0.50–0.59	5
0.60–0.69	0.3
0.70–0.79	0.6
0.80–0.89	0.3
0.90–0.99	0
1.00–1.19	0
1.20–1.39	0
1.40–1.59	0
1.60–1.79	0

Source: Lewis, 1951.

Table 5

Frequencies of Occurrence (in %) of Different Median Droplet Diameters, Total Droplet
Concentrations, and Average Droplet Diameters in Stratus and Stratocumulus Clouds

Droplet Diameter Range (μm)	Pacific Coast (60 cases)	Other Parts of the USA (267 cases)
0–9	5	32
10–14	36	43
15–19	25	16
20–24	17	6
25–29	7	2
> 29	10	1
Droplet Concentration (cm^{-3})	100	320
Average Droplet Diameter (μm)	19.8	11.6

Source: Lewis, 1951.

Table 6

Temperature Dependence of the Liquid Water Content in Stratus, Stratocumulus, and Altocumulus Clouds over the Soviet Union

Temperature intervals (in ° C)	Number of observations	Average liquid water content $(g\ m^{-3})$	Maximum liquid water content $(g\ m^{-3})$
-35.0, -30.1	1	0.15	0.15
-30.0, -25.1	3	0.09	0.15
-25.0, -20.1	47	0.12	0.34
-20.0, -15.1	163	0.13	0.82
-15.0, -10.1	710	0.14	1.47
-10.0, -5.1	1334	0.18	0.99
-5.0, -0.1	1542	0.21	1.53
0.0, 4.9	663	0.30	3.00
5.0, 9.9	301	0.33	3.14
10.0, 14.9	108	0.27	1.18
15.0, 19.9	15	0.29	1.14

Source: Borovikov et al., 1963.

where it was as much as 50% subadiabatic. The concentration of droplets remained relatively constant throughout the cloud, deviating by only about 10%, and the mean droplet size increased monotonically with height. The measurements also clearly showed that cloud-top entrainment produced a decrease in liquid water content and droplet concentration.

From measurements in Arctic stratus clouds, Herman and Curry (1984) found a wide range of cloud liquid-water contents, increasing on average from cloud base to cloud top. The effective droplet radius (ratio of the volume of droplets to their surface area) r_e increased from about 4 to 5 μm near cloud base to 8 to 10 μm just below cloud top. The peak droplet concentrations were of the order of 100 cm^{-3}.

B. Stratocumulus (Sc)

The physical dimensions and microphysical properties of stratocumulus clouds are listed in Table 3 and Tables 4–7, respectively. Maritime stratocumulus is the most common type of stratus clouds, yet it has not been characterized as extensively as continental stratocumulus. Stratocumulus in maritime areas is generally characterized by warm base temperatures (5 to 10° C) and is typically several

Table 7

Microphysical Structures of Stratocumulus Clouds

Reference	Total droplet Concentration (cm^{-3}) [a]	Liquid water content $(g\ m^{-3})$	Mean droplet volume diameter (μm)	Location/ comments
Albrecht et al. (1985)	~25-125	~0.10	~18	Off California coast. From 3 flights. Ratio of measured to adiabatic liquid water content 0.4 to 0.8 at cloud base; 0.1 to 0.4 at cloud top.
Brost et al. (1982a,b)	—	<0.10	—	Off California coast. From 5 flights.
Caughey & Kitchen (1984)	~200–500	~0.15	~10	Over Great Britain. 2 Case Studies. Ratio of measured to adiabatic liquid water content ~0.9 at cloud base, 0.4 at cloud top.
Hobbs & Rangno (1985)	302	—	—	Over Washington state & Pacific Ocean. 17 cases.
Nicholls (1984)	~200	~0.7 max	~12 near cloud base ~20 near cloud top	Over the North Sea. 1 flight.
Paluch & Lenschow (1991)	—	~0.2	—	Off California coast. 10 Flights. Measurements may be too low.
Slingo et al. (1982)	200–400	~0.5 max.	8 near cloud base 14 near cloud top	Over England. 3 Cases.

[a] Above about 2 μm diameter.

hundred meters thick (Brost et al., 1982*a, b*). There is often an inversion close to cloud base. The evolution of maritime stratocumulus can be strongly influenced by a capping inversion generated by large-scale subsidence over a cool ocean, as well as by entrainment of dry air. Stratocumulus clouds over land areas commonly have lower base temperatures than maritime stratocumulus.

The discussion here will focus on findings mostly from case studies in maritime stratocumulus. Droplet size spectra measured through a stratocumulus cloud layer have revealed that cloud droplets increase in size with height, as would be expected from classical droplet growth equations, and larger droplets precipitate through the base (Nicholls, 1984). In these clouds, the measured droplet concentrations remained nearly constant and the mean volume radius increased from cloud base to near cloud top. The liquid water content was subadiabatic. Nicholls also showed that while there was evidence of coalescence producing droplets of at least 200 μm radius, such drops contributed only a few percentages to the total liquid water content.

Caughey and Kitchen (1984) reported an association between larger droplets near cloud base and the presence of colder plumes, suggesting that the microphysical structures of marine stratocumulus clouds may be modulated by convective motions. Roach et al. (1982) and Slingo et al. (1982) investigated the radiative and microphysical properties of stratocumulus over Great Britain using balloon data. Albrecht et al. (1985) showed that microphysical processes may affect the structure of stratocumulus clouds as well as their precipitation production, and that cloud structure is important in regulating the distribution of radiative cooling in clouds, at least those off the coast of California.

Twomey and Cocks (1989), Rawlins and Foot (1990), and Nakajima et al. (1991), by comparing effective droplet radii estimated from remote radiometric sensors to nearly coincident droplet size spectra measured *in situ* in stratocumulus, demonstrated that remote sensors consistently overestimate the radii of cloud droplets by 2 to 3 μm. The differences have been attributed to uncertainties in the treatment of cloud reflectance in solar wavelengths.

C. Droplet Size Distributions

A wealth of aircraft measurements in the Soviet Union indicate that droplet size spectra in stratocumulus are distributed in logarithmic normal (Levin, 1954) or gamma forms (Borovikov et al., 1963). The droplet size spectra in stratus and stratocumulus are now commonly described by the gamma distribution:

$$n(r) = \frac{N_0}{\Gamma(\alpha + 1)\beta^{\alpha+1}} r^\alpha \exp\left(\frac{-r}{\beta}\right) = \frac{N_0}{\Gamma(\alpha + 1)\beta^{\alpha+1}} r^\alpha \exp\left(\frac{3r}{r_{av}}\right) \quad (1)$$

where r is the droplet radius; r_{av} is the average radius (very close to the true arithmetic mean); $n(r)$ is the droplet concentration (cm^{-3}), Γ is the incomplete gamma distribution (Sekhon and Srivastava, 1970; Gradshteyn and Ryzhik, 1980); and N_0, α, and β are constants. The constant α is found from the average radius, and N_0 and β are derived from curve fits. For clouds of a given type the droplet size spectra, averaged over large spatial scales as well as over many cases, are typically well described by a particular form of Eq. (1) with $\alpha = 2$ (the so-called Khrgian–Mazin distribution). Table 8 summarizes some relationships for describ-

Table 8

Parameters Derived from Gamma-Sized Distribution

Total droplet concentration: $\int\limits_{0}^{\infty} n(r)dr$

Liquid water content: $\frac{4\pi}{3}\rho_w \int\limits_{0}^{\infty} r^3 n(r)dr$

where, ρ_w is the density of water.

Modal radius: $r_{mod} = \beta\alpha$

Volume modal radius: $r_{v_{mod}} = \beta(\alpha + 3)$

Total droplet cross-section: $S = \pi N_o \beta^2 (\alpha + 1)(\alpha + 2)$

Water content: $Q = \frac{4}{3}\pi N_o \rho_w \beta^3 (\alpha + 1)(\alpha + 2)(\alpha + 3)$

Standard deviation: $\sigma_r = \beta(\alpha + 1)^{1/2}$

Source: Mazin and Khrgian, 1989.

ing cloud droplet characteristics and a number of the important size spectrum characteristics.

A normalized droplet size spectrum can be described by

$$f(r) = \frac{1}{\Gamma(\alpha + 1)\beta^{\alpha+1}} r^\alpha \exp\left(\frac{-r}{\beta}\right) \tag{2}$$

It can also be shown that

$$\frac{N_0}{\Gamma(\alpha + 1)\beta^{\alpha+1}} = \frac{1.45 \text{LWC}}{r_{av}^6} \times 10^{-6} \tag{3}$$

where LWC is the liquid water content (g m^{-3}). This equation does not reproduce the droplet spectral form for $r \gtrsim 20$ μm; Mazin and Khrgian (1989) discuss means of computing the distribution above this size. For stratus and stratocumulus clouds in the range from 0.1 to 0.6 km above cloud base, the following expression is given by Mazin and Khrgian (1989):

$$\bar{r} = 3.4(1 + b\omega)(1 + 0.3a_1 z + 0.3a_2 z^2 + 0.3a_3 z^3) \tag{4}$$

where $\bar{r}$ is an average mean droplet radius (in μm); temperature at cloud base is ω in °C; height z above cloud base is in kilometers; and $a_1 = 5.0$ km^{-1}, $a_2 = -6.0$ km^{-2}, $a_3 = 2.1$ km^{-3}; and $b = 3.7 \times 10^{-2}$ (°C)$^{-1}$.

According to Mazin and Khrgian (1989), in stratus and stratocumulus clouds with thickness H up to 600–700 m the liquid water content increases with height almost up to the cloud top and then decreases rapidly. Mazin and Khrgian (1989) derived the following analytical expression for stratus and stratocumulus when the cloud thickness exceeds 700 m:

$$\overline{LWC} = A(1 + b\omega)(1 + a_1 z + a_2 z^2 + a_3 z^3) \tag{5}$$

where $\overline{LWC}$ is "mean" liquid water content; ω and z are defined in the previous paragraph. The coefficients in Eq. (5) in stratus and stratocumulus are as follows: $A = 0.13$ g m^{-3}, $b = 3.7 \times 10^{-2}$ (°C)$^{-1}$, $a_1 = 5.0$ km^{-1}, $a_2 = -6.0$ km^{-2}, and $a_3 = 2.1$ km^{-3}. Equation (5) can also be applied to cumulus clouds, but different coefficients are needed (Mazin and Khrgian, 1989).

By substituting $\bar{r}$ in Eq. (4) for r_{av} in Eq. (1) as a first approximation, and using $\overline{LWC}$ from Eq. (5) for LWC in Eq. (3), $n(r)$ and the related parameters that describe a droplet spectrum in Table 8 can be derived.

IV. Mixed-Phase Clouds

A. Occurrence of Supercooled Liquid Water

The relative quantities of liquid water and ice in a cloud at a temperature below 0°C will depend on the temperature; the age of the cloud; the ambient aerosol concentrations, composition, and size; and possibly other factors.

A climatology of liquid water occurrence as a function of temperature was compiled from a large data set collected in the Soviet Union from 1953–1958 in a variety of cloud types (see Fig. 1). The probability of finding liquid water dimin-

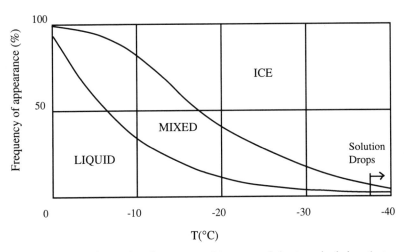

Figure 1   Average frequencies of appearance of the supercooled water, mixed-phase (water and ice), and all ice as a function of temperature in layer clouds over the European territory of the USSR. (After Borovikov et al., 1963.)

ishes from 100% at 0°C to 0% at about −40°C (the homogeneous ice nucleation point). The lowest temperatures at which clouds consisted only of liquid water was found to be −36.3°C.

Heymsfield and Sabin (1989) tabulated results on the detection of liquid water in continental mid- and upper-level clouds (As, Ac, Ci) as a function of temperature from −20° to −50°C. A total of more than 90,000 data points, each representing 1 second of data or 90 m of flight path, were used in the compilation. Liquid water was measured only 1% of the time and was not detected below −35.3° C. Sassen and Dodd (1988) reported on a case study in orographic wave clouds that showed that micrometer-sized drops freeze "homogeneously" at temperatures below about −37°C. A systematic study of the presence of liquid-water in orographically generated lenticular wave clouds was conducted by Heymsfield and Miloshevich (1993). They used a Rosemount icing probe with a liquid water detection threshold of approximately 0.002 g m^{-3} and, as an independent verification, a cryogenic hygrometer that provided a reliable indication of whether water-saturated conditions were present. Micrometer-sized drops or larger were not present below −37°C, even within 10 m s^{-1} updrafts; below this temperature water saturation is not reached since the ice crystals that form are numerous enough to rapidly reduce the relative humidity to ice saturation. However, these authors present measurements that strongly suggest submicrometer-sized solution droplets are present at least down to −40°C.

B.   Altostratus (As) and Altocumulus (Ac)

Altostratus clouds appear as gray or bluish sheets or layers, ranging from several hundreds to thousands of meters thick, often covering the entire sky. The layers are variously striated, fibrous, or uniform in aspect. Altocumulus clouds form in the mid-troposphere, in distinct layers often less than 100 m thick. From the ground, altocumulus clouds are identified by their sharp outlines, enclosing rounded, often convective elements. Their outlines reflect their tendency toward liquid water composition (Huschke, 1970).

Although prevalent in the atmosphere and potentially important in the earth's radiation budget, the microphysical and radiative properties of altocumulus have not received much attention. Few, if any, systematic studies of the temporal evolution of the ice phase in these clouds have been conducted. Rather, the limited amount of work on the topic has centered around defining their microphysical properties. Summaries of the measured properties of these clouds appear in Tables 9–11.

Borovikov et al. (1963) reported droplet diameters averaging about 10 μm from *in situ* measurements in altostratus and altocumulus. Herman and Curry (1984), in a detailed description of the microstructure of a 300 meter deep altocumulus sampled between −9 and −12°C, reported mean droplet diameters be-

Table 9

Microphysical Structures of Altostratus and Altocumulus Clouds

Reference	Droplet concentration (cm^{-3})[a]	Liquid water content (g m^{-3})	Mean droplet volume diameter (μm)	Location/ comments
Borovikov et al. (1963)		0.086 to 0.17	9–11	Soviet Union.
Herman & Curry (1984)	300	0.03–0.09	7–10	Alaska. 1 Case. -9 to -12 °C.
Heymsfield et al. (1990)	100	<0.007	<9	Wisconsin. 11 cloud penetrations. 100 m thick. -30 ° C. Little ice.
Heymsfield et al. (1991)	25	<0.10	15–20	Wisconsin. 2 clouds. Top -30 ° C. Little Ice.
Hobbs & Rangno (1985)	256 range: (70–1000)	—	—	Over or near Washington State. 31 cases. Cloud tops -5 to -26 °C.
Lewis (1951)	35–75	most <0.20	14–19	USA. 240 cases. See Tables.

[a] Above about 2 μm droplet diameter.

tween 7 and 10 μm, droplet concentrations averaging about 300 cm^{-3}, and liquid water contents from 0.03 to 0.09 g m^{-3}. Hobbs and Rangno (1985) investigated the microstructure of 31 altocumulus of different genus (species), ranging in thickness from 100 to 1000 m, most commonly 200 to 500 m. Cloud-top temperatures were between -5 and -26°C with all but two of the clouds warmer than -20°C. Droplet concentrations averaged from 70 to 1000 cm^{-3}, with peak liquid water contents of 0.1 g m^{-3}. Typically, the maximum ice particle concentration in these clouds was much less than 10 per liter, and very often no ice particles were detected. The microstructure of several thin (100 m) altocumulus clouds at

Table 10

Microphysical Properties of Altostratus and
Altocumulus Clouds

Liquid water content (g m^{-3})	A_c, A_c–A_s, (% of 246 observations)
0.00–0.09	50
0.10–0.19	32
0.20–0.29	13
0.30–0.39	4
0.40–0.49	1
0.50–0.59	0
0.60–0.69	0
0.70–0.79	0
0.80–0.89	0
0.90–0.99	0
1.00–1.19	0
1.20–1.39	0
1.40–1.59	0
1.60–1.79	0

Source: Lewis, 1951.

about $-30°C$ was described by Heymsfield et al. (1990). The mean droplet diameters were less than 9 μm, the peak droplet concentrations less than 100 cm^{-3}, and the mean liquid water content 0.007 g m^{-3} or less. Virtually no ice particles were detected. An investigation of a different altocumulus cloud at about $-30°C$ showed droplet concentrations of only about 25 cm^{-3}, but with mean droplet diameters as large as 15 to 20 μm and very little ice (Heymsfield et al., 1991).

The size spectra of droplets in altostratus and altocumulus clouds have been given in the references listed in Table 9; these spectra are generally narrow compared with those in the lower-level stratus and stratocumulus clouds. Droplet size spectra in altostratus and altocumulus as a function of temperature and cloud thickness are given by Eq. (4), according to Mazin and Khrgian (1989), and the same coefficents as for stratus and stratocumulus are applicable. Ice particle concentrations, habits, and sizes in altostratus and altocumulus are highly variable.

V. Ice Clouds

In this section we will discuss the structure of cirrus clouds, the most prevalent type of ice cloud. According to the Glossary of Meteorology (Huschke, 1970), cirrus are detached clouds in the form of white, delicate filaments or mostly white

Table 11

Frequencies of Occurrence (in %) of Different Average Effective Droplet
Diameters, Total Droplet Concentration, and Average Droplet Diameter
in Altocumulus and Altostratus Clouds

Average effective droplet diameter range (μm)	Pacific coast (112 cases)	Other parts of the USA (128 cases)
0–9	8	20
10–14	22	32
15–19	28	30
20–24	22	12
25–29	7	5
> 29	13	1
Droplet Concentration (cm^{-3})	35	75
Average Droplet Diameter (μm)	18.8	14.2

Source: Lewis, 1951.

patches or narrow bands, composed almost exclusively of ice crystals. Dense patches or tufts of cirrus may contain ice crystals large enough to acquire appreciable fallspeeds; extended trails of virga may sometimes be seen below such clouds. Wind shear and particle size affect the shape of the trails. One cirrus type displaying these features is cirrus uncinus; it is comma-shaped, terminating at the top in a hook or tuft, which is presumably an ice generation region. The definition of cirrus given by the World Meteorological Organization (1956) is extended to include the glaciated anvils of thunderstorms far from the active convective clouds that produce them, but excluding the cirruslike, stratiform clouds associated with mesoscale convective complexes, and low-level ice clouds and ice fogs near the poles. As noted earlier, supercooled water droplets are occasionally found at temperatures as low as -36 or $-37°C$, presumably when the clouds are in their formative stages or in association with vigorous convection.

Cirrus clouds frequently occur in layers or sheets covering hundreds or even thousands of kilometers. Because their horizontal dimensions are much greater than their vertical extent, they are referred to as stratiform or layered clouds. Although such terminology suggests a statically stable lapse rate, thin layers of conditional or convective instability often exist and are conducive to the formation of small-scale convection embedded in such clouds.

Observational data on cirrus cloud thicknesses and heights have been reviewed mostly for the United States by Stone (1957) and Dowling and Radke (1990); by Borovikov et al. (1963) and Kosarev and Mazin (1991) primarily for the Soviet Union; and by Platt and Dilley (1981) for Melbourne, Australia (40°S latitude). Dowling and Radke reported a mean cloud thickness of about 1.5 km. The mean thickness of the clouds studied by Platt and Dilley was 2.6 km. The data in Kosarev and Mazin suggest that the cloud thickness is dependent on latitude, with the peak frequency at several hundred meters for mid- and low-latitude cirrus, and at about 1.2 km in high-latitude regions. According to Borovikov et al. and Kosarev and Mazin, cloud-top heights decrease with increasing latitude (Table 12) in a manner consistent with the change in height of the tropopause. In the mid- and high latitudes, cloud-base heights were from 5.5 to 8 km, and in the lower latitudes base heights ranged from 10 to 14 km. Dowling and Radke reported that cloud-top heights were approximately three quarters of the height of the tropopause: about 9 km over the continental United States and 13 km near the equator. The mean cloud-top height measured by Platt and Dilley at Melbourne was 9.8 km. Because lidar is now being used in a systematic way to investigate the physical dimensions and frequency of occurrence of cirrus clouds (Cox et al., 1987; Sassen

Table 12

Cirrus Heights

Location	Latitude	Altitude of Ci – Cs (km)	
		average	maximum
Bossekop (Northern Norway)	70°N	7.3	
Pavlovsk (USSR) – Uppsala (Sweden)	60°N	7.6	
Potsdam (West Germany) – Trappe (France)	51°N	8.7	12.67
Blue Hill – Washington (USA)	40°N	10.15	15.01
Mera (Japan)	35°N	11.02	16.79
Manila (Philippines)	14°N	12.05	20.45
Jakarta (Indonesia)	6°S	11.04	18.60

Source: Borovikov et al., 1963.

et al., 1990) a more complete cloud-climatology of these properties of cirrus should be available within a few years.

A.   Microphysics of Cirrus Forming *In Situ*

An explanation of the distinctive, fibrous structure of cirrus-generating cells, sometimes referred to as cirrus uncinus, was given by Ludlam (1947), who suggested that the curdled appearance near the top of most cirrus clouds was evidence for water droplets. He also concluded that the trails seen below them were ice particle precipitation. In the 1950s, a radar group from Canada's McGill University conducted extensive investigation of generating cells. The results were published in a series of papers (Marshall et al., 1952; Douglas et al., 1957). Other radar investigations conducted by Plank et al. (1955) and Braham and Spyers-Duran (1967) showed that cirrus ice crystals could fall into the top of liquid water clouds, quite efficiently converting liquid water to ice and subsequent precipitation.

Weickmann (1947) conducted some early airborne investigations into the microstructure of cirrus clouds. From ice crystals collected *in situ*, he found that the primary crystalline form in cirrocumulus, cirrus spissatus, and cirrus fibratus were hollow crystals, or three-dimensional clusters of prismatic crystals joined at a common center ("bullet rosettes"). He reported that the crystals in cirrostratus were primarily individual hexagonal columns and plates. Recent observations in cirrus indicate that ice particle habits are dependent on the temperature (Heymsfield and Platt, 1984). At temperatures above about $-45°C$, ice particles are complex, three-dimensional forms; at lower temperatures they are in geometrically more simple forms of hexagonal columns and plates (Heymsfield and Platt, 1984). Particle habit is almost certainly related to the ambient relative humidity as well as temperature; but it has only recently been possible to measure relative humidity at low temperatures, at least *in situ* (Oltmans, 1986; Spyers-Duran and Schanot, 1987). Cirrus particles can evidently aggregate, producing many particles larger than 300 μm (Kajikawa and Heymsfield, 1989).

Characteristics of cirrus ice particle size spectra and associated size spectral moments have been reported in a number of studies (e.g., Heymsfield and Knollenberg, 1972; Hobbs et al., 1974; Heymsfield, 1975; Griffith et al., 1980; Heymsfield and Platt, 1984; Sassen et al., 1989; and Heymsfield et al., 1990). The results have been presented in summary form by Liou (1986), Dowling and Radke (1990), and Jeck (1988).

The measured particle size spectra in cirrus are indicative of the decreases in the amount of water vapor available for growth as the temperature decreases. The shape of the size spectra decreases exponentially or in a power-law form with increasing size, and the slope of the size spectra steepens with decreasing temperature (Heymsfield and Platt, 1984). The largest particles in a given size spectrum increase with increasing temperature: When temperatures exceed $-40°C$,

Table 13

Ice Water Content (in grams per cubic meter)[a]

Top left

Temp Interval(°C)	Mean	Sigma	Number of points	Maximum	Median	75%ile
-90,-80	0.00000	0.00000	0	0.00000	0.00000	0.00000
-80,-70	0.00000	0.00000	0	0.00000	0.00000	0.00000
-70,-60	0.00000	0.00000	0	0.00000	0.00000	0.00000
-60,-50	0.00076	0.00036	5	0.00130	0.00049	0.00077
-50,-40	0.00477	0.00786	16	0.03109	0.00187	0.00442
-40,-30	0.05785	0.05936	23	0.26864	0.03408	0.05308
-30,-20	0.19673	0.13813	14	0.39778	0.15000	0.20527
-20,-10	0.65287	0.50960	7	1.61000	0.38583	0.38583
-10,0	0.00000	0.00000	0	0.00000	0.00000	0.00000

$\ln(IWC)=0.558+0.0667T-0.0016T^2$, $(r^2=0.99; -15>T(°C)>-55)$

Top right

Temp Interval(°C)	Mean	Sigma	Number of points	Maximum	Median	75%ile
-90,-80	0.00013	0.00002	5	0.00015	0.00011	0.00012
-80,-70	0.00397	0.00000	1	0.00397	0.00397	0.00397
-70,-60	0.00309	0.00294	222	0.01829	0.00273	0.00407
-60,-50	0.00468	0.00649	415	0.08141	0.00232	0.00676
-50,-40	0.01376	0.03084	382	0.18270	0.00237	0.00719
-40,-30	0.04749	0.07144	726	0.77220	0.01143	0.07119
-30,-20	0.07925	0.07426	686	0.39120	0.05872	0.12060
-20,-10	0.13852	0.12870	561	0.78350	0.09728	0.17090
-10,0	0.23029	0.17592	149	1.24100	0.17970	0.23610

$\ln(IWC)=-1.20+0.0896T+0.000126T^2$, $(r^2=0.92; -5>T(°C)>-85)$

Bottom left

Temp Interval(°C)	Mean	Sigma	Number of points	Maximum	Median	75%ile
-90,-80	0.00000	0.00000	0	0.00000	0.00000	0.00000
-80,-70	0.00000	0.00000	0	0.00000	0.00000	0.00000
-70,-60	0.00000	0.00000	0	0.00000	0.00000	0.00000
-60,-50	0.00000	0.00000	0	0.00000	0.00000	0.00000
-50,-40	0.00735	0.00481	124	0.02140	0.00870	0.01010
-40,-30	0.00859	0.00480	794	0.01760	0.01010	0.01250
-30,-20	0.01289	0.00372	963	0.02130	0.01010	0.01510
-20,-10	0.01221	0.00476	311	0.02210	0.01380	0.01520
-10,0	0.00000	0.00000	0	0.00000	0.00000	0.00000

$\ln(IWC)=-4.25-0.0044T-0.00036T^2$, $(r^2=0.97; -15>T(°C)>-45)$

Bottom right

Temp Interval(°C)	Mean	Sigma	Number of points	Maximum	Median	75%ile
-90,-80	0.00000	0.00000	0	0.00000	0.00000	0.00000
-80,-70	0.00000	0.00000	0	0.00000	0.00000	0.00000
-70,-60	0.00000	0.00000	0	0.00000	0.00000	0.00000
-60,-50	0.00000	0.00000	0	0.00000	0.00000	0.00000
-50,-40	0.07194	0.03462	18	0.14700	0.05300	0.07800
-40,-30	0.30532	0.29253	367	1.41800	0.19700	0.39900
-30,-20	0.44511	0.45914	231	2.49600	0.36000	0.50700
-20,-10	0.64369	0.61408	117	2.87300	0.48000	0.82100
-10,0	0.00000	0.00000	0	0.00000	0.00000	0.00000

$\ln(IWC)=-1.40-0.0817T-0.0026T^2$, $(r^2=0.99; -15>T(°C)>-45)$

[a]Summary of ice water content measurements as a function of temperature in cirrus clouds forming *in situ* and from deep convection in tropical cirrus and in mid-latitude continental cirrus. Top left: Environmental Definition Program (EDP), cirrus and cirrostratus over the continental United States. (Data points are averages over about 20 km horizontal distances.) Bottom left: FIRE data in cirrus and cirrostratus over Wisconsin, U.S.A. (Averages over about 1 km intervals.) Top right: Tropical cirrus (*in situ* and anvil) over Kwajalein, Marshall Islands. (Averages over about 1 km intervals.) Bottom right: Data from anvil cirrus from the CCOPE program in Montana, U.S.A. (Averages over about 1 km intervals.) The number of data points (N) and the mean, maximum, median, and 75 percentile are shown. Fitted relationships to the mean ice water content (IWC) values as a function of temperature (T, in °C) are also shown.

the largest particles often exceed 1 mm in length, while below about $-60°C$ they are at most 10 to 100 μm.

The measurements also indicate that ice crystal concentrations in cirrus fall in the range 0.01 to 1.0 cm^{-3}, most typically from 0.01 to 0.1 cm^{-3}, and in general the concentrations decrease with decreasing temperature. However, the magnitude of the total concentrations of the size spectra are uncertain because of the inability to *reliably* measure ice crystal size spectra below about 25 μm in length (Brown, 1989; Dowling and Radke, 1990; Wielicki et al., 1990; Heymsfield et al., 1990; Ackerman et al., 1990). There is evidence of high concentrations of ice crystals with length less than about 25 μm, especially at temperatures below about $-45°C$, or when particles form in an ice-free region. For example, Platt et al. (1989) found evidence for the presence of ice particles 10 μm in length or smaller in cirrus clouds below a temperature of about $-60°C$; Heymsfield (1986*a*) reported a similar observation in tropical cirrus at about $-80°C$; and Heymsfield and Miloshevich (1993) measured ice crystal concentrations in excess of 100 cm^{-3} in orographic lenticular wave clouds at temperatures of $-37°C$ and below.

The measurements in continental cirrus formed *in situ* indicate the expected decrease in ice water content (IWC) with decreasing temperature, from as large as 0.1 g m^{-3} above $-40°C$ to about 10^{-4} g m^{-3} below about $-60°C$; examples from two field programs that took samples within ice clouds are shown in Table 13, along with empirical relationships fitted to the data. The two data sets differ at the higher temperatures because the cirrus in the deeper case ("EDP"—see Heymsfield, 1977) merged into middle and lower-level clouds, while in the shallower case (the First ISCCP Regional Experiment, FIRE—Heymsfield et al., 1990) cirrus base and top were well-defined. The EDP data represent averages over about 20 km horizontal distance while the FIRE data span a distance of about 0.5 km.

For further information, the reader is referred to Table 3 in Dowling and Radke (1990), which includes all of the measurements reported in the above-referenced articles, except those since 1990. The reader is also referred to a special issue of the Monthly Weather Review (Vol. 118, No. 11, 1990) for a comprehensive study of cirrus clouds, carried out as part of FIRE.

B. Anvil Cirrus

Our knowledge of the microphysics of anvil cirrus is less complete than that of cirrus formed *in situ*. This is because anvils, particularly their upper regions, are not as accessible to aircraft; where they are accessible the ice crystals are often too small to be measured with available instrumentation. Although the distinction is arbitrary, in the discussion that follows anvil cirrus are distinguished from precipitating, stratiform regions of convective clouds or convective complexes.

A few measurements on the microphysical structures of anvil cirrus in maritime tropical areas have been reported. For example, particle habits near the tropopause are bullet rosettes (Knollenberg et al., 1982), and ice water contents vary from about 10^{-4} g m^{-3} near the tropopause (Knollenberg et al., 1982) to several tenths of a gram per cubic meter at mid-anvil levels (Griffith et al., 1980).

A summary of previously unreported IWC measurements at Kwajalein in the Marshall Islands (8°N latitude), in a combination of anvil cirrus (removed from the convective cores) and cirrus formed *in situ*, are presented as a function of temperature in Table 13, along with empirical relationships fitted to the data. Heymsfield (1986*a*) discusses the data in the tropical cirrus below −80°C. These data represent averages over about 20-km distances. The data points are plotted in Fig. 2.

Bennetts and Ouldridge (1984) have described aircraft measurements in the anvil of a winter maritime cumulonimbus. They reported that crystal concentrations occasionally exceeded 0.15 cm^{-3} and that the ice particle sizes were rather uniformly divided, with half of the mass contained in particles larger than 1 mm in diameter.

There is less information on the microphysical structure of anvils forming in association with mid-latitude, continental convective clouds. In a study combining satellite and radar data, Heymsfield et al. (1983) measured lower infrared temperatures at the edges of such an anvil than in its interior. This suggests that both ice water content and ice crystal concentrations are greater at the anvil boundary. Evidence exists for the formation of particles as large as 1 cm diameter at temperatures as low as −40°C; these resulted from the aggregation of single, smaller ice particles within the anvils. This evidence comes from data collected in Montana thunderstorm anvils (Heymsfield, 1986*b*). Ice water contents are higher than cirrus studied in other locations [see Table 13 and the curve fits, based on Heymsfield (1986*b*), where each point represents an average over about 10 km]. Furthermore, the ice particles in these anvil clouds are large in comparison with particles generated in cirrus forming *in situ*. Obviously, the ice content of anvil cirrus will depend on the intensity of the storm that produces it, the position downwind of the storm, and the environmental wind shear.

VI. Some Outstanding Problems

This chapter has provided a review of the physical dimensions and microphysical structure of those cloud types that are likely to have the greatest effect on climate—namely, stratus and stratocumulus, altostratus and altocumulus, and cirrus. The following topics need to be addressed before the effects of these clouds on climate can be evaluated in a realistic manner:

- Quantitative and measurable criteria for defining cloud types and characteristics of their structure.
- Cloud condensation nucleus concentrations as a function of geographical location and altitude.

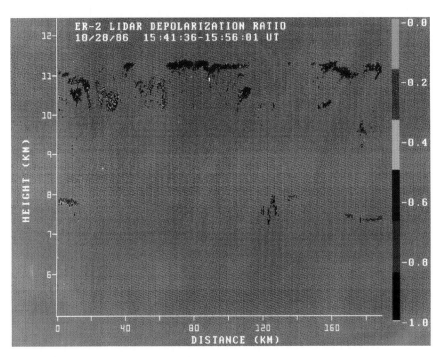

Plate 1.   Lidar depolarization ratio as a function of altitude and distance along the nadir track of the ER-2 on 28 October 1986. The depolarization ratio for the upper cloud layer and the lower diffuse cloud at 120 km is large, due primarily to the presence of ice crystals. The thin altocumulus layer having a low depolarization ratio consists primarily of supercooled water droplets. (From Spinhirne and Hart, 1990.)

Longwave Radiation
1985-1986

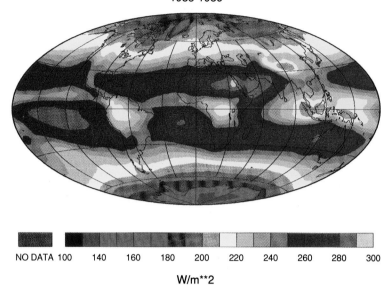

NO DATA 100   140   160   180   200   220   240   260   280   300

W/m**2

Longwave Cloud Forcing
1985-1986

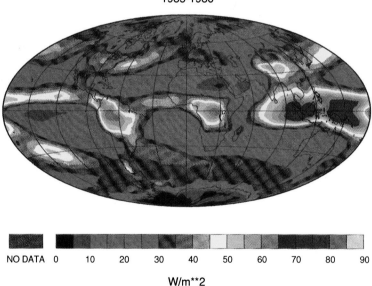

NO DATA 0   10   20   30   40   50   60   70   80   90

W/m**2

Plate 2.   Annual average outgoing longwave radiation (top) and longwave cloud radiative forcing (bottom) determined from two years (1985–1986) of ERBE scanner data on the ERBS and NOAA-9 satellites.

Absorbed Solar Radiation
1985-1986

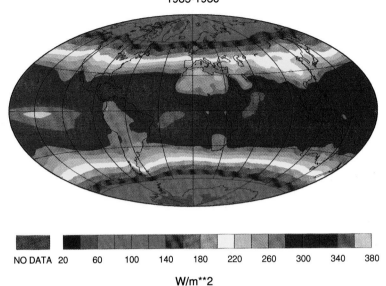

NO DATA 20   60   100   140   180   220   260   300   340   380

W/m**2

Shortwave Cloud Forcing
1985-1986

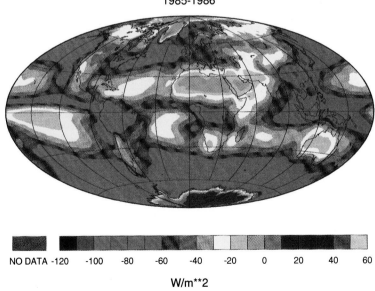

NO DATA -120   -100   -80   -60   -40   -20   0   20   40   60

W/m**2

Plate 3.   Annual average absorbed solar radiation (top) and shortwave cloud forcing (bottom) determined from two years (1985–1986) of ERBE scanner data on the ERBS and NOAA-9 satellites.

Net Radiation
1985-1986

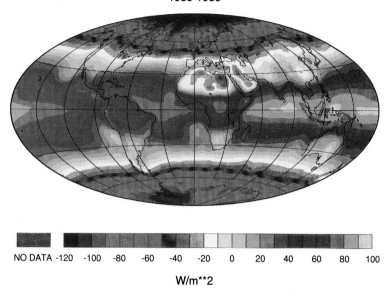

NO DATA -120 -100 -80 -60 -40 -20 0 20 40 60 80 100

W/m**2

Net Cloud Forcing
1985-1986

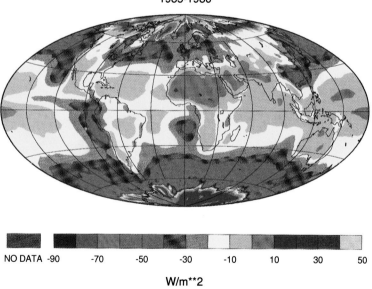

NO DATA -90 -70 -50 -30 -10 10 30 50

W/m**2

Plate 4. Annual average net radiation (top) and net cloud radiative forcing (bottom) determined from two years (1985–1986) of ERBE scanner data on the ERBS and NOAA-9 satellites.

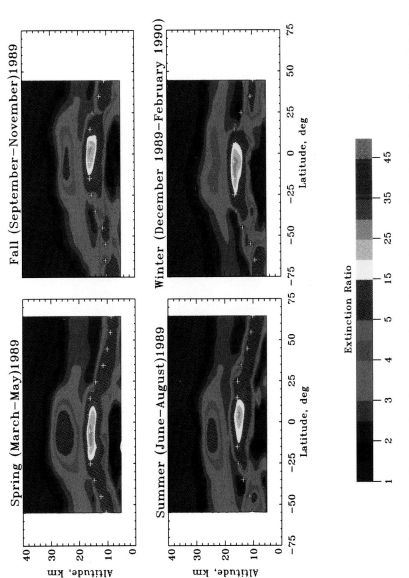

Plate 5. Seasonally averaged zonal mean aerosol extinction ratio (i.e., ratio of aerosol extinction coefficient to molecular extinction coefficient) for 1989 northern hemisphere spring (March–May), summer (June–August), fall (September–November), and winter (December 1989–February 1990). Measurements from SAGE II.

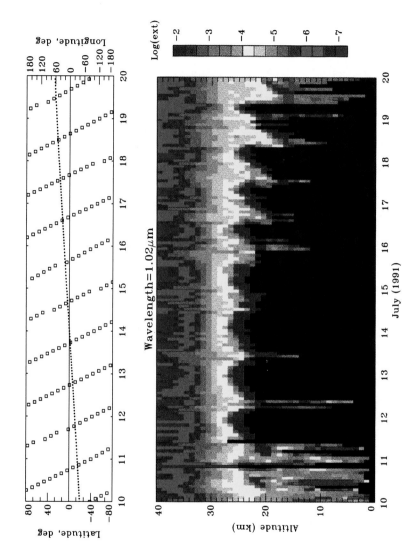

Plate 6. Time series of the SAGE II aerosol extinction profile measurements at 1.02-μm wavelength for July 10–20, 1991. Dots represent latitude; solid squares represent longitude.

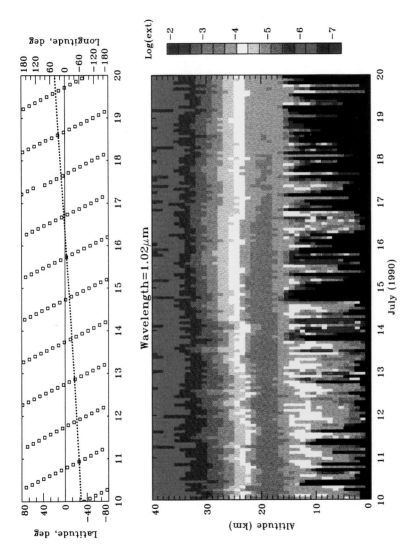

Plate 7. Time series of the SAGE II aerosol extinction profile measurements at 1.02-μm wavelength for July 10–20, 1990. Dots represent latitude; solid squares represent longitude.

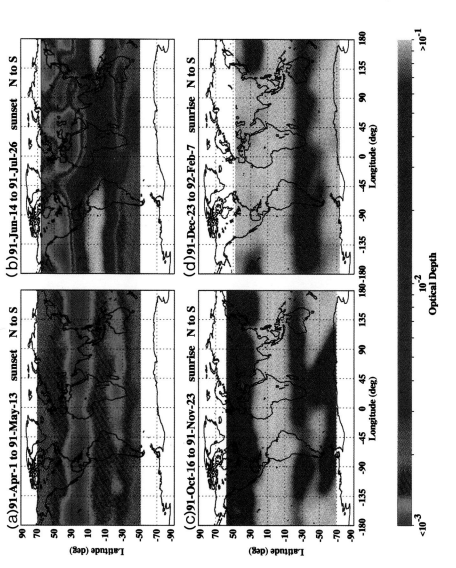

Plate 8. Longitude–latitude distributions of SAGE II 1.02-μm stratospheric aerosol optical depth: (a) April 1, 1991, to May 13, 1991; (b) June 14, 1991, to July 26, 1991; (c) October 16, 1991, to November 23, 1991; and (d) December 23, 1991, to February 7, 1992.

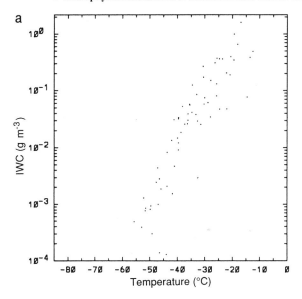

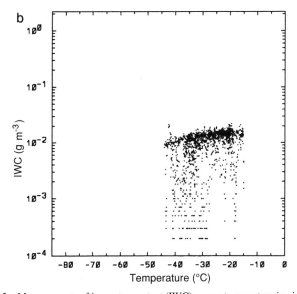

Figure 2   Measurements of ice water content (IWC) versus temperature in cirrus formed *in situ* in thunderstorm anvils. Continental cirrus (EDP) (a) and continental cirrus (FIRE) (b). The experiments and parameters shown are explained in Table 13.

(*continues*)

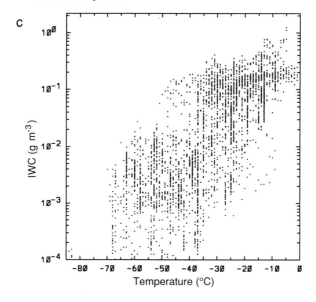

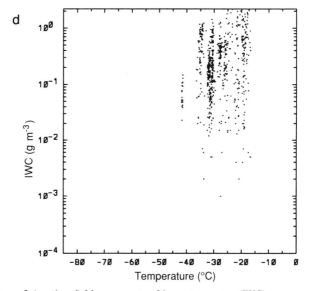

Figure 2 (*continued*) Measurements of ice water content (IWC) versus temperature formed *in situ* in thunderstorm anvils. Tropical cirrus (Kwajalein) (c) and continental anvil cirrus (CCOPE) (d). The experiments and parameters shown are explained in Table 13.

- Droplet size distributions and liquid water contents in stratus and stratocumulus for a wide range of geographical areas.
- Relative humidity distribution in the upper troposphere.
- Primary and secondary ice nucleation mechanisms and ice nucleus concentrations and composition as a function of temperature and height in the atmosphere.
- Morphology (shape) of cirrus ice crystals as a function of vapor density and temperature.
- Cirrus ice crystal size distributions, in particular below 25 μm.
- Knowledge of the scattering behavior (phase function) of ice crystals, especially for the complex crystals found at temperatures below $-20°C$.
- Mean cloud bulk properties (i.e., $\overline{IWC}$, optical depth) as a function of horizontal and vertical average dimensions.

Many of these topics could be addressed with aircraft that are better suited for long-duration flights in the uppermost parts of the troposphere, improved aircraft instrumentation for measuring cloud microphysical properties, and reliable means for characterizing the properties of clouds from remote sensors, especially satellite radiometers.

Acknowledgments

The author wishes to thank Nancy Knight and Professor Peter Hobbs for their contributions to this manuscript. The author also wishes to thank Greg Melvin for the preparation of this manuscript.

References

Ackerman, S. A., Smith, W. L., Spinhirne, J. D., and Revercomb, H. E. (1990). *Mon. Wea. Rev.* **118**, 2377–2388.
Albrecht, B. A., Penc, R. S., and Schubert, W. H. (1985). *J. Atmos. Sci.* **42**, 800–822.
Albrecht, B. A., Fairall, C. W., Thomson, D. W., White, A. B., Snider, J. B., and Schubert, W. H. (1990). *Geophys. Res. Letters.* **17**, 89–92.
Beard, K. V. (1987). *Review of Geophysics* **25**, 357–370.
Bennetts, D. A., and Ouldridge, M. (1984). *Quart. J. Roy. Meteor. Soc.* **110**, 85–103.
Borovikov, A. M., Gaivoronskii, L. I., Zak, E. G., Kostarev, V. V., Mazin, I. P., Minervin, V. E., Khrgian, A. Kh., and Shmeter, S. M. (1963). *Cloud Physics*. Israel Program for Scientific Translations, Jerusalem, Israel, 392 pp.
Braham, R. R. Jr., and Spyers-Duran, P. (1967). *J. Appl. Met.* **6**, 1053–1061.
Brost, R. A., Lenschow, D. H., and Wyngaard, J. C. (1982a). *J. Atmos. Sci.* **39**, 800–817.
Brost, R. A., Lenschow, D. H., and Wyngaard, J. C. (1982b). *J. Atmos. Sci.* **39**, 818–836.
Brown, P. A. (1989). *J. A. O. Tech.* **6**, 293–306.
Caughey, S. J., and Kitchen, M. (1984). *Quart. J. Roy. Meteor. Soc.* **110**, 13–34.
Cooper, W. A. (1991). *Reviews of Geophysics, Supplement* **29**, 69–79.

Cox, S. K., McDougal, D. S., Randall, D. A., and Schiffer, R. A. (1987). *Bull. Amer. Meteor. Soc.* **68**, 114–118.

Douglas, R. H., Gunn, K. L. S., and Marshall, J. S. (1957). *J. Met.* **14**, 95–108.

Foot, J. S. (1988). *Quart. J. Roy. Meteor. Soc.* **114**, 129–144.

Dowling, D. R., and Radke, L. F. (1990). *J. Appl. Meteor.* **29**, 970–978.

Gradshteyn, I. S., and Ryzhik, I. M. (1980). Table of Integrals, Series, and Products. Orlando, 1160 pp.

Griffith, K. T., Cox, S. K., and Knollenberg, R. G. (1980). *J. Atmos. Sci.* **37**, 1077–1087.

Herman, G. F., and Curry, J. A. (1984). *J. Atmos. Sci.* **23**, 5–24.

Heymsfield, A. J. (1975). *J. Atmos. Sci.* **32**, 799–808.

Heymsfield, A. J. (1977). *J. Atmos. Sci.* **35**, 284–295.

Heymsfield, A. J. (1986a). *J. Atmos. Sci.* **43**, 851–855.

Heymsfield, A. J. (1986b). *J. Atmos. Sci.* **43**, 2463–2478.

Heymsfield, A. J., and Knollenberg, R. G. (1972). *J. Atmos. Sci.* **29**, 1358–1366.

Heymsfield, A. J., and Platt, C. M. R. (1984). *J. Atmos. Sci.* **41**, 846–855.

Heymsfield, A. J., and Sabin, R. M. (1989). *J. Atmos. Sci.* **46**, 2252–2264.

Heymsfield, A. J., Miller, K. M., and Spinhirne, J. D. (1990). *Mon. Wea. Rev.* **118**, 2313–2328.

Heymsfield, A. J., Miloshevich, L. M., Slingo, A., Sassen, K., and Starr, D. O'C. (1991). *J. Atmos. Sci.* **48**, 923–945.

Heymsfield, A.J., and Miloshevich, L. M. (1993). *J. Atmos. Sci.* (in press).

Heymsfield, G. M., Szejwach, G., Schotz S., and Blackmer, R. H., Jr. (1983). *J. Atmos. Sci.* **40**, 1756–1767.

Hobbs, P. V., Chang, S., and Locatelli, J. D. (1974). *J. Geophys. Res.* **79**, 2199–2206.

Hobbs, P. V., and Rangno, A. L. (1985). *J. Atmos. Sci.* **42**, 2523–2549.

Hudson, J. G. (1983). *J. Atmos. Sci.* **40**, 480–486.

Huschke, R. E. (1970). *Glossary of Meteorology.* American Meteorological Society, Boston, Mass., 638 pp.

Jeck, R. K. (1988). Federal Aviation Administration report DOT/FAA/CT-89/3.

Kajikawa, M., and Heymsfield, A. J. (1989). *J. Atmos. Sci.* **46**, 3108–3121.

Knollenberg, R. G., Dascher, A. J., and Huffman, D. (1982). *Geophys. Res. Lett.* **9**, 613–616.

Kosarev, A. L., and Mazin, I. P. (1991). *Atmos. Res.* **26**, 213–228.

Levin, L. M. (1954). *Dokl. Akad. Nauk. SSSR* **94f**, 1045–1053.

Lewis, W. (1951). In *Compendium of Meteorology* Published by the American Meteorological Society, 1197–1203.

Liou, K. N. (1986). *Mon. Wea. Rev.* **114**, 1167–1199.

Ludlam, F. H. (1947). II. *Quart. J. Roy. Meteor. Soc.* **73**, 257–265.

Marshall, J. S., Langleben, M. P., and Rigby, E. Caroline (1952). *McGill University Stormy Weather Group* Scientific Report **MW-8**, 23pp.

Mazin, I. P., and Khrgian, A. Kh. (1989). Handbook of Clouds and Cloudy Atmosphere. Leningrad, 646 pp.

Nakajima, T., King, M. D., Spinhirne, J. D., and Radke, L. F. (1991). *J. Atmos. Sci.* **48**, 728–750.

Nicholls, S. (1984). *Quart. J. R. Meteo. Soc.* **110**, 783–820.

Nicholls, S., and Leighton, J. (1986). *Quart. J. Roy. Meteor. Soc.* **112**, 431–460.

Noonkester, V. R. (1984). *J. Atmos. Sci.* **41**, 829–845.

Oltmans, S. J. (1986). NOAA Data Rep. ERL-ARL-7.

Plank, V. G., Atlas, D., and Paulsen, W. H. (1955). *J. Meteor.* **12**, 358–378.

Platt, C. M. R., and Dilley, A. C. (1981). *J. Atmos. Sci.* **38**, 671–678.

Platt, C. M. R., Spinhirne, J. D., and Hart, W. D. (1989). *J. Geophy. Res.* **94**, 11,151–11,164.

Pruppacher, H. R., and Klett, J. D. (1978). *Microphysics of Clouds and Precipitation.* Reidel, Dordrecht, Holland, 714 pp.

Ramanathan, V., Pitcher, E. J., Malone, R. C., and Blackmon, M. L. (1983). *J. Atmos. Sci.* **40**, 605–630.

Ramanathan, V., and Collins, W. (1991). *Nature* **351**, 27–32.

Rawlins, F., and Foot, J. S. (1990). *J. Atmos. Sci.* **47**, 2488–2503.

Roach, W. T., Brown, R., Caughey, S. J., Grease, B. A., and Slingo, A. (1982). *Quart. J. Roy. Meteor. Soc.* **108**, 103–132.

Rogers, R. R., and Yau, M. K. (1989). *A short course in cloud physics.* Pergamon Press, 209 pp.

Sassen, K., and Dodd, G. C. (1988). *J. Atmos. Sci.* **45**, 1357–1369.

Sassen, K., and Dodd, G. C. (1989). *J. Atmos. Sci.* **46**, 3005–3014.

Sassen, K., Starr, D. O'C. and Uttal, T. (1989). *J. Atmos. Sci.* **46**, 371–396.

Sassen, K., Grund, C. J., Spinhirne, J. D., Hardesty, M. M., and Alvarez, J. M. (1990). *Mon. Wea. Rev.* **118**, 2288–2311.

Sekhon, R. S., and Srivastava, R. C. (1970). *J. Atmos. Sci.* **27**, 299–307.

Slingo, A., Brown, R., and Wrench, C. L. (1982). *Quart. J. Roy. Meteor. Soc.* **108**, 145–166

Spyers-Duran, P., and Schanot, A. (1987). *Proc. AMS Sixth Symposium on Meteorological Observations and Instrumentation, New Orleans, La.,* 209–212.

Stone, R. G. (1957). *Air Weather Service Technical Report AWSTR105–130,* 156 pp.

Twomey, S., and Cocks, T. (1989). *Beitr. Phys. Atmos.* **62**, 172–179.

Warren, S. G., Hahn, C. J., London, J., Chervin R. M., and Jenne, R. (1986). NCAR Tech. Note TN-273 STR, 229 pp.

Warren, S. G., Hahn, C. J., London, J., Chervin R. M., and Jenne, R. (1988). NCAR Tech. Note TN-317 STR, 212 pp.

Weickmann, H. (1947). *Die Eisphase in der Atmosphare.* **Library Trans. 273,** Royal Aircraft Establishment, 96 pp.

Wielicki, B. A., Suttles, J. T., Heymsfield, A. J., Welch, R. M., Spinhirne, J. D., Wu, M. C., Starr, D. O., Parker L., and Arduini, R. F. (1990). *Mon. Wea. Rev.* **118**, 2356–2376.

World Meteorological Organization, 1956. *Internat. Cloud Atlas* **Vols. 1 and 2.**, published by The World Meteorological Organization, Geneva.

Chapter 5 | Radiative Properties of Clouds

Michael D. King
NASA Goddard Space Flight Center
Greenbelt, Maryland

This paper presents an overview of our current understanding of the radiative properties of clouds, placing particular emphasis on recent results and unanswered problems arising from the marine stratocumulus and cirrus cloud components of the First ISCCP (International Satellite Cloud Climatology Project) Regional Experiment (FIRE), conducted in the United States during 1986 and 1987. For marine stratocumulus clouds, we present and discuss the discrepancy between observations and theory of the absorption of solar radiation by clouds, the discrepancy between remote sensing and *in situ* estimates of the effective droplet radius derived from spectral reflectance measurements, and the variability and spatial structure of stratocumulus clouds derived from both reflection and transmission measurements. We will describe the thermal emission characteristics of cirrus clouds and will demonstrate how the brightness temperature difference in the split-window region of the thermal infrared can be used to infer the effective radius of ice crystals, observations that lead to the conclusion that ice crystals are much smaller than previously believed. We will also illustrate the relationship between thermal emittance and visible albedo that has been derived from airborne observations of cirrus clouds. These results generally show that the thermal emittance of cirrus clouds is less than theoretically predicted for a given value of the visible albedo.

I. Introduction

Clouds vary considerably in their horizontal and vertical extent (Stowe et al., 1989; Rossow et al., 1989), in part due to the circulation pattern of the atmosphere with its requisite updrafts and downdrafts, and in part due to the distribution of oceans and continents and their numerous and varied sources of cloud condensation nuclei (CCN). Clouds strongly modulate the energy balance of the earth and its atmosphere through their interaction with shortwave and longwave radiation, as demonstrated both from satellite observations (Ramanathan, 1987; Ramanathan et al., 1989) and from modeling studies (Ramanathan et al., 1983; Cess et al., 1989). Of paramount importance to a comprehension of the earth's climate and its response to anthropogenic and natural variability is a knowledge of the radiative, microphysical, and optical properties of clouds.

Marine stratocumulus clouds exert a large influence on the radiation balance of the earth–atmosphere–ocean system through their large aerial extent, temporal persistence, and high reflectivity of solar radiation. Cirrus clouds, on the other hand, exert their greatest radiative influence on the earth's climate through their effect on longwave radiation emitted to space. Both of these cloud types are spatially and temporally persistent in the earth's atmosphere, and both create difficulty in the remote sensing of cloud properties from spaceborne sensors. As a direct consequence of the need to determine the optical and microphysical properties of clouds from present and future spaceborne systems, such as the Moderate Resolution Imaging Spectroradiometer (MODIS; King et al., 1992), a need arose to conduct intensive field observations (IFOs) of marine stratocumulus and cirrus clouds. These two field campaigns, conducted as major components of the First ISCCP Regional Experiment (FIRE; Cox et al., 1987), itself an element of the International Satellite Cloud Climatology Project (ISCCP; Schiffer and Rossow, 1983), has focused exclusively on these two cloud types. Largely as a result of these two field experiments, the radiative and microphysical properties of these cloud systems have been more extensively studied than others.

In this chapter, we summarize the state of our knowledge of the radiative properties of clouds based on these and other experiments that have had an especially profound impact on our understanding of cloud radiative properties. We begin this review by examining the principal observations that have contributed to our knowledge of the absorption of solar radiation by clouds, placing particular emphasis on recent observations and explanations for the widely observed discrepancy between theory and observations, wherein clouds are often observed to absorb more solar radiation than models can explain. In addition, we will describe the status of a number of recent efforts to determine the microphysical and radiative properties of clouds from reflected solar radiation measurements, again focusing on marine stratocumulus observations.

We will also illustrate the spatial variability of the spectral reflectance and the angular transmission characteristics of clouds based on aircraft observations conducted off the coast of Southern California during the marine stratocumulus IFO. For cirrus clouds, we will summarize recent findings on the thermal emission characteristics of these clouds, together with a description of the effective radius of ice crystals inferred from aircraft and satellite observations. Finally, we will describe the relationship between the thermal emittance and visible albedo of cirrus clouds derived from actual field observations.

II.   Marine Stratocumulus Clouds

A.   Spectral Absorption of Solar Radiation

The absorption of solar radiation by clouds is governed by the optical thickness, single scattering albedo, and phase function of the cloud, as well as the reflectance of the underlying surface and the water vapor distribution of the environment in which the cloud is located. Theoretical calculations suggest that water clouds absorb up to 15–20% of the incident solar radiation, with the largest values arising from the thickest clouds having large cloud droplets, an overhead sun, and little water vapor above the cloud (Twomey, 1976; Slingo and Schrecker, 1982; Davies et al., 1984; Stephens et al., 1984; Wiscombe et al., 1984; Slingo, 1989). In addition to the total cloud absorption, calculations also show that heating rates near cloud top can reach 2 K h^{-1}, thereby contributing significantly to the sudden "burning off" of California stratus layers as the solar zenith angle decreases toward noon (Twomey, 1983).

The majority of cloud absorption observations to date have involved measurements obtained using broadband pyranometers mounted on research aircraft flown above and below clouds. All of the observations thus far reported in the literature have involved single aircraft missions in which it is exceedingly difficult to obtain comparable flux observations above and below the same cloud layer. In spite of these difficulties, aircraft pyranometer observations by Reynolds et al. (1975), Herman (1977), Stephens et al. (1978), Herman and Curry (1984), Hignett (1987), and Foot (1988) have consistently shown a discrepancy between measurements and theory, whereby measurements of the absorption of solar radiation by clouds generally suggest that clouds absorb more solar energy than theoretical predictions can explain for clouds composed solely of liquid water and water vapor.

As a consequence of the reported discrepancies between measurements and theory on the absorption of solar radiation by clouds, recognized by the atmospheric radiation community for more than four decades, a large number of competing hypotheses have been offered to explain this "anomalous absorption paradox." Each of these hypotheses has a somewhat different ramification for the spectral distribution of the excess absorption between the visible and near-infrared portions of the solar spectrum and, as such, cannot be resolved using standard broadband pyranometer measurements. For example, Twomey (1972, 1977) suggested that absorbing aerosol particles, either within the cloud droplets or interstitial to them, may be partly responsible for this excess absorption. Calculations by Newiger and Bähnke (1981) show that absorbing aerosol particles interstitial to the cloud droplets can enhance cloud absorption to values up to 30% of the incident solar radiation, but that this effect is largely restricted to wavelengths $\lambda \lesssim 1.5$ μm. The possibility also exists that leakage of radiation through the sides of clouds might account for some of the large values of absorption implied by the measurements (Welch et al., 1980; Ackerman and Cox, 1981). Wiscombe et al.

(1984) suggested that significant concentrations of large, drizzle-sized droplets could contribute to larger absorption values than typically obtained from calculations, but this effect would be restricted largely to wavelengths $\lambda \gtrsim 1.5$ μm. Finally, Stephens and Tsay (1990) suggested that an unobserved water-vapor continuum, if found to be present, might contribute to explaining this "anomalous absorption paradox."

In response to the widely recognized limitation of single-aircraft broadband pyranometer measurements, together with the need to spectrally resolve the absorption measurements in order to distinguish between the various competing anomalous absorption hypotheses, we were prompted to develop the diffusion domain method for determining the absorption of solar radiation by clouds as a function of wavelength (King, 1981). In this method, the intensity of scattered radiation deep within a cloud layer is measured as a function of zenith angle for selected wavelengths in the visible and near-infrared.

Figure 1 is a schematic illustration of the various regimes of an optically thick cloud layer illuminated from above by solar radiation incident at solar zenith angle θ_0. Deep within an optically thick medium, located sufficiently far from the top and bottom boundaries of the medium (cloud), the diffuse radiation field assumes an asymptotic form characterized by rather simple properties. In this region, known as the diffusion domain, the role of direct (unscattered) radiation is negligible compared to the role of diffuse radiation, the diffuse intensity field is azimuthally independent, and the relative angular distribution is independent of solar zenith angle. For a vertically homogeneous cloud layer at a wavelength for which the single scattering albedo $\omega_0 < 1$, the intensity in the diffusion domain is given by

$$I(\tau,\ u) = s_1 P(u)e^{-k\tau} + s_2 P(-u)e^{-k(\tau_c - \tau)} \tag{1}$$

where τ is the optical depth measured from the upper boundary of the cloud; τ_c is the total optical thickness of the cloud layer; u is the cosine of the zenith angle with respect to the positive τ direction ($-1 \leq u \leq 1$); $P(u)$ is the diffusion pattern (eigenfunction); k is the diffusion exponent (eigenvalue); and s_1 and s_2 are the strengths of the diffusion streams in the positive and negative τ directions, respectively (see Fig. 1). As the relative strengths between s_1 and s_2 depend only on τ_c and the optical properties of the cloud layer, the relative angular distribution of the diffuse radiation field is independent of solar zenith angle.

The transition between the upper- and lowermost levels of the cloud layer and the innermost diffusion domain occurs at a scaled optical depth $(1 - g)\tau \simeq 2$ from both the top and bottom boundaries of the cloud, where g is the asymmetry factor. This transition is shown schematically in Fig. 1 and numerically in Herman et al. (1980), but it is only approximate in that the transition is a gradual one and the precise level of the transition depends on the accuracy required of the asymptotic formulas applicable in the diffusion regime.

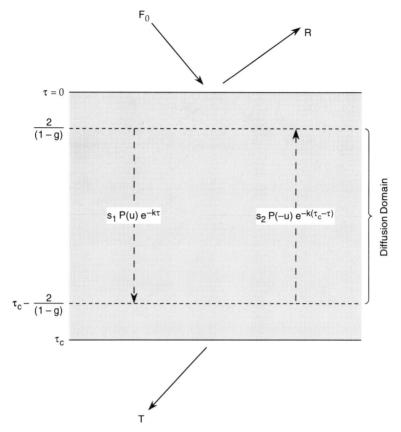

Figure 1 Schematic illustration of the radiation regimes of an optically thick, nonconservative cloud layer illuminated from above by solar radiation incident at a solar zenith angle θ_0. The diffusion domain, located deep within the cloud and sufficiently far from its top and bottom boundaries, is a regime characterized by azimuthally symmetric radiation that can be characterized by a sum of upward- and downward-propagating diffusion streams.

Figure 2 illustrates the relative intensity as a function of zenith angle for various values of the similarity parameter (s), where s is a function of the cloud asymmetry factor (g) and single scattering albedo (ω_0) as follows:

$$s = \left(\frac{1 - \omega_0}{1 - \omega_0 g} \right)^{1/2} \qquad (2)$$

The zenith angle is here defined with respect to the downward-directed normal such that $\theta = 0°$ (180°) corresponds to a zenith (nadir) measurement or, alternatively, to nadir (zenith) propagating radiation. This figure pertains to computations

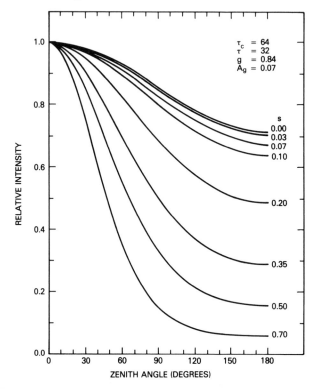

Figure 2   Relative intensity as a function of zenith angle and similarity parameter at the midlevel of a cloud of total optical thickness $\tau_c = 64$. These curves apply to a Henyey-Greenstein phase function with $g = 0.84$ and when the surface reflectance $A_g = 0.07$. (From King et al., 1986.)

made for an optical depth $\tau = 32$ in a cloud of total optical thickness $\tau_c = 64$ when the surface reflectance $A_g = 0.07$. These calculations, based on the phase function introduced by Henyey and Greenstein (1941) for $g = 0.84$, clearly show that the diffuse intensity field in the diffusion domain of an optically thick cloud layer is expected to monotonically decrease from zenith to nadir. Figure 2 shows computations of the internal scattered radiation field for similarity parameters ranging between 0.0 and 0.7 which, for an asymmetry factor $g = 0.84$, corresponds to $0.87 \leq \omega_0 \leq 1.0$.

Mel'nikova (1978) was the first to suggest that the ratio of the upward to downward propagating fluxes within the diffusion domain be used to determine the single scattering albedo of clouds. King (1981) demonstrated that the ratio of the nadir-to-zenith intensities within a cloud layer is far more sensitive to single scattering albedo than is the ratio of the upward-to-downward propagating fluxes, and further showed analytically that this ratio is a function solely of A_g, s, and the

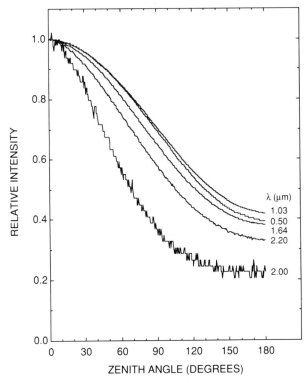

Figure 3 Relative intensity as a function of zenith angle and wavelength for internal scattered radiation measurements obtained with the cloud absorption radiometer on 10 July 1987. (From King et al., 1990.)

scaled optical depth between the aircraft flight level and the base of the cloud $[(1 - g)(\tau_c - \tau)]$.

The first experimental observations of the angular distribution of scattered radiation deep within a cloud layer, together with an analysis of the spectral similarity parameter derived from these measurements, were presented and discussed by King et al. (1990). The data were obtained using the cloud absorption radiometer (CAR) described by King et al. (1986), which flew on the University of Washington's Convair C-131A aircraft during the FIRE marine stratocumulus IFO, conducted off the coast of San Diego, California during July 1987. The microphysical structure of the clouds, including the cloud droplet size distribution, was also monitored continuously with instruments aboard the aircraft.

Figure 3 shows the relative intensity as a function of zenith angle obtained from measurements inside clouds for selected wavelengths of the cloud absorption radiometer. Aside from the quantization (digitization) noise in the shortest wavelength channels and instrumental (electrical) noise in the longest wavelength channels, two main features can be seen on examination of Fig. 3. These are (1) the angular intensity field at the shortest wavelength follows very nearly the cosine function expected for conservative scattering in the diffusion domain, and (2) the angular intensity field becomes increasingly anisotropic as absorption increases. This is especially noticeable at 2.0 μm, the instrument channel where water droplets have their greatest absorption. The experimental observations presented in Fig. 3 complement the theoretical figure (for selected values of the similarity parameter rather than wavelength) presented in Fig. 2.

Finally, making use of the analytic formulations inherent in the diffusion domain method, together with measurements obtained with the cloud absorption radiometer, King et al. (1990) derived the spectral similarity parameter for clouds in a 50-km section of marine stratocumulus located some 355 km southwest of the airfield on Coronado Island, San Diego. These results, presented in Fig. 4, illustrate the mean and standard deviation of the spectral similarity parameter for all 13 channels of the CAR obtained on 10 July 1987. Although the conversion

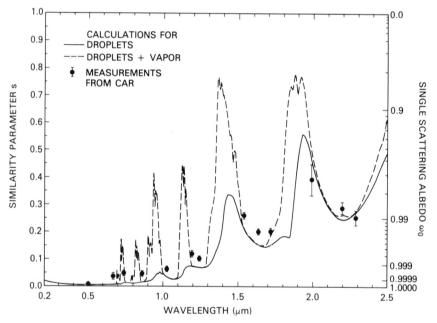

Figure 4   Calculations of the similarity parameter s as a function of wavelength for water droplets alone (solid line) and drops plus vapor (dashed line) for the cloud droplet size distribution and water vapor conditions of the marine stratocumulus cloud of 10 July 1987. The single scattering albedo scale is valid at $\lambda = 0.754$ μm, where the cloud asymmetry factor $g = 0.848$. The measurements derived from the cloud absorption radiometer (solid circles with error bars) are averages of the similarity parameter derived by applying the diffusion domain method to a 50-km section of this cloud. (From King et al., 1990.)

from s to ω_0 is not unique, due to the moderate spectral variation of g, a single scattering albedo scale has been provided in this figure as a matter of convenience. This scale, shown on the right-hand side of Fig. 4, is strictly applicable at $\lambda = 0.754$ μm. Based on profile ascents and descents following these measurements, the stratocumulus cloud layer was determined to be 440 m thick with a cloud base at approximately 490 m.

In addition to the experimental results obtained using the CAR, Fig. 4 illustrates calculations of the similarity parameter as a function of wavelength for a cloud composed of water droplets only (solid curve) and droplets plus saturated vapor at 10.3°C (dashed curve). The water droplet computations were based on calculations for the measured cloud droplet size distribution. The water-vapor computations, on the other hand, were based on assuming the cloud was composed of saturated vapor having a column loading of water vapor of 0.45 g cm^{-2}. The water-vapor transmission functions were computed for this cloud layer at a resolution of 20 cm^{-1} using LOWTRAN 5 (Kneizys et al., 1980). The absorption optical depths thus obtained were combined with the corresponding optical properties for cloud droplets, where the total optical thickness of the cloud was assumed to be 16 at a wavelength of 0.754 μm. The total optical thickness of this 50-km section of cloud was approximately 32.3 $\pm$ 4.2, which affects the relative weighting between cloud droplets and water vapor. We recalculated the theoretical curves in Fig. 4 using LOWTRAN 7 (Kneizys et al., 1988) and $\tau_c = 16$ and 32, but these differences have only a minor impact on the conclusions drawn from this figure.

The results presented in Fig. 4 show that, *in this case*, the measured absorption of solar radiation by clouds is close to, but persistently larger than, theoretical calculations. Furthermore, these findings support the view that clouds absorb more and reflect less solar radiation than theoretical predictions. It was not possible to bring theory and measurements into complete agreement by simply postulating an error in the measurement of the effective radius (r_e), as this would improve the agreement in some parts of the spectrum and worsen the agreement in other parts of the spectrum.

The close agreement between measurements and theory in this case, where measurements show a small but consistently larger absorption than theoretical predictions, is consistent with modest "anomalous absorption" in these clouds that were largely free of anthropogenic influence. The single scattering albedos that we obtained from our analysis in the visible wavelength region, though somewhat lower than theory, are still $\sim$0.9999, values generally much too large to explain any reduced reflection by these clouds ($20 \lesssim \tau_c \lesssim 42$). On the other hand, in the wavelength region between 1.6 and 2.2 μm, our measurements of excess absorption are consistent with the observations of Twomey and Cocks (1982), Stephens and Platt (1987), and Foot (1988), who reported unusually low spectral reflectance in this wavelength region. Furthermore, recent changes in computations of water vapor absorption properties within both the absorption bands and window regions of the near-infrared, as reflected in LOWTRAN 7 (Kneizys et al., 1988), suggest that the theoretical calculations in the presence of water vapor (dashed line in Fig. 4) may have to be modified to some degree. Thus, it appears that a combination of new measurement techniques, new instruments, and revisions in our theoretical treatment of water vapor absorption in light of new measurements of line parameters are all promising new advances that are leading toward a solution of this four-decade-old anomaly in the shortwave absorption by terrestrial water clouds.

B. Cloud Optical Thickness and Effective Particle Radius

A number of efforts have been devoted to determining the cloud optical thickness and/or effective particle radius from reflected solar radiation measurements, both from aircraft (Hansen and Pollack, 1970; Twomey and Cocks, 1982, 1989; King, 1987; Foot, 1988; Rawlins and Foot, 1990; Nakajima et al., 1991) and satellite (Curran and Wu, 1982; Arking and Childs, 1985; Rossow et al., 1989) platforms. In each of these methods, multiwavelength radiometers have been used to obtain measurements of the reflection function $R(\tau_c; \mu, \mu_0, \phi)$, formed from a ratio of the reflected intensity $I(0, -\mu, \phi)$ and the incident solar flux F_0 as follows:

$$R(\tau_c; \mu, \mu_0, \phi) = \frac{\pi I(0, -\mu, \phi)}{\mu_0 F_0} \tag{3}$$

In this expression μ_0 is the cosine of the solar zenith angle θ_0; μ is the absolute value of the cosine of the zenith angle θ, measured with respect to the positive τ direction; and ϕ is the relative azimuth angle between the direction of propagation of the emerging radiation and the incident solar direction.

Radiative transfer theory shows that the reflection function of optically thick layers is largely a function of the scaled optical thickness $\tau_c' = (1 - \omega_0 g)\tau_c$ and the similarity parameter s, where the similarity parameter, in turn, depends primarily on the effective particle radius, defined by (Hansen and Travis, 1974)

$$r_e = \int_0^\infty r^3 n(r) \; dr \; / \int_0^\infty r^2 n(r) \; dr \tag{4}$$

where $n(r)$ is the particle size distribution and r is the particle radius. In addition to τ_c', s, and A_g, the details of the single scattering phase function affect the directional reflectance of the cloud layer (King, 1987).

The fundamental principle behind the simultaneous determination of the cloud optical thickness and effective particle radius is that the reflection function of clouds at a weakly absorbing channel in the visible wavelength region is primarily a function of the cloud optical thickness, whereas the reflection function at a water (or ice) absorbing channel in the near-infrared is primarily a function of cloud particle size. This can most easily be seen on examination of Fig. 5, which shows simultaneous computations of the reflection function of clouds at 0.75 and

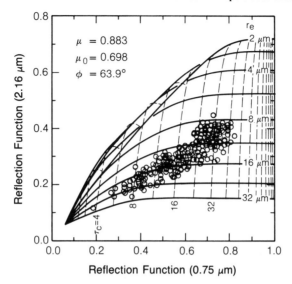

Figure 5 Theoretical relationship between the reflection function at 0.75 and 2.16 μm for various values of the cloud optical thickness (at λ = 0.75 μm) and effective particle radius for the case when θ_0 = 45.7°, θ = 28.0°, and ϕ = 63.9°. Data from measurements above marine stratocumulus clouds during FIRE are superimposed on the figure. (From Nakajima and King, 1990.)

2.16 μm for various values of τ_c and r_e when θ_0 = 45.7°, θ = 28.0°, and ϕ = 63.9°. These angles correspond to a case for which observations were obtained of the reflectance of marine stratocumulus clouds during FIRE in July 1987. The data points superimposed on this figure were obtained from the NASA ER-2 aircraft using the multispectral cloud radiometer (MCR) described by Curran et al. (1981) and King (1987), from which we conclude that this 145-km section of cloud had an optical thickness at 0.75 μm that ranged between 6 and 45 with an effective radius that ranged between 8 and 22 μm.

Whether one formulates the retrieval of r_e in terms of a ratio of the reflection function at two wavelengths, as in Foot (1988), Twomey and Cocks (1989), and Rawlins and Foot (1990), or as an absolute reflection function, as in Curran and Wu (1982) and Nakajima and King (1990), the underlying physical principles behind the retrieval remain the same. As the effective radius increases, absorption monotonically increases for all $r_e \gtrsim 1$ μm. As a consequence, cloud reflectance in the near-infrared (e.g., 2.16 μm) generally decreases (see Fig. 5). At a weakly absorbing channel in the visible wavelength region (e.g., 0.75 μm), on the other hand, the reflection function depends primarily on the total optical thickness such that τ_c increases as the the reflection function increases.

During the marine stratocumulus IFO, a major effort was expended in obtaining data sets of both the spectral reflection and microphysical properties of clouds to enable these remote sensing concepts to be validated. Figure 6 compares the

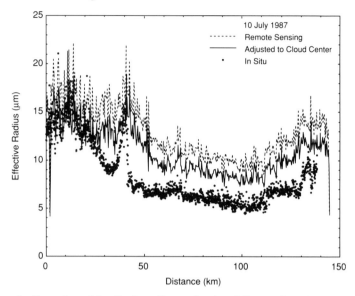

Figure 6   Comparison of the effective radius as a function of distance along the nadir track of the ER-2 aircraft derived from remote sensing (dashed line) and *in situ* measurements from the C-131A aircraft (solid circles). The solid line represents the expected values of effective radius at the geometric center of the cloud layer, derived from the remote-sensing measurements by allowing for vertical inhomogeneity of droplet radius. (From Nakajima et al., 1991.)

retrieved effective radius with *in situ* estimates obtained along the nadir track of the ER-2 aircraft. The remote sensing values of the effective radius (dashed curve) were adjusted to the expected values at the geometric center of the cloud layer (solid curve) using the method outlined by Nakajima and King (1990), thereby taking into account the vertical distribution of microphysical properties typical of marine stratocumulus clouds. This case study (after Nakajima et al., 1991) is unique in that the ER-2 aircraft was well coordinated with the University of Washington C-131A aircraft, which was making nearly simultaneous *in situ* microphysical measurements (solid circles). These results clearly show that the shape of the retrieved r_e time series is generally similar to, but systematically larger than, the direct *in situ* measurements of the effective radius. These results suggest, therefore, that clouds reflect less solar radiation at 2.16 μm than theoretically predicted (see Fig. 5), which is likewise consistent with cloud absorption being enhanced over theory in this "window" region (see Fig. 4).

A similar conclusion was reached by Twomey and Cocks (1989), who measured the spectral reflectance of marine stratus off the coast of eastern South Australia and subsequently measured the droplet size distribution from the same aircraft platform. Figure 7 is a histogram of the effective radius derived from their five-channel remote sensing algorithm and from *in situ* microphysical measurements. One again draws the conclusion that remote sensing overestimates the ef-

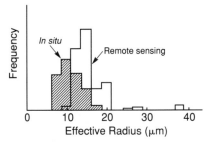

Figure 7 Histogram plot of the frequency distribution of inferred effective radius for one day's observations as derived by remote sensing (open histogram) and as derived from *in situ* microphysical measurements (hatched). (From Twomey and Cocks, 1989.)

fective radius in comparison with *in situ* measurements. To clarify the bias in the retrieved values of r_e, Twomey and Cocks (1989) presented cumulative distributions of r_e derived from both remote sensing and *in situ* measurements. These results, presented in Figs. 8a (logarithmic scale) and 8b (linear scale), clearly show that remote sensing overestimated the effective radius by ~5 μm (~40%), and was more like an offset (bias) than a percentage overestimation (similar to the conclusions of Nakajima et al., 1991).

Rawlins and Foot (1990) utilized the United Kingdom C-130 aircraft during the FIRE marine stratocumulus experiment to determine τ_c and r_e from remote sensing measurements. They utilized both reflection and transmission function measurements, and in both cases found their retrieved values of r_e to be in excess of *in situ* microphysical measurements by some 2–3 μm (25–50%). In addition, Rawlins and Foot (1990) obtained better agreement in r_e for the optically thinner parts of the cloud, as did Nakajima et al. (1991). This is contrary to expectations, in that one would expect a retrieval algorithm based on spectral reflectance mea-

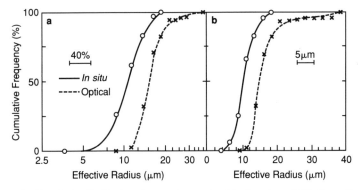

Figure 8 Cumulative distribution of remotely sensed and *in situ* values of effective radius plotted on (a) a logarithmic scale, and (b) a linear scale. (From Twomey and Cocks, 1989.)

surements to be more accurate for optically thick clouds due to the enhanced sensitivity to absorption arising from increased multiple scattering. Both Rawlins and Foot (1990) and Nakajima et al. (1991) obtained satisfactory estimates of the cloud optical thickness (not shown), further suggesting that the bulk of the discrepancy in the retrieval of τ_c and r_e is largely confined to the optical properties of clouds in the near-infrared (e.g., 2.16 μm).

Although some discrepancy still remains between *in situ* and remote sensing estimates of r_e, the bulk of recent evidence suggests that the excess absorption by clouds is largely restricted to the water-vapor window regions of the near-infrared. Numerous studies have demonstrated that clouds absorb more and reflect less solar radiation than theoretically predicted. Stephens and Tsay (1989) were the first to suggest that an overlooked absorption by water vapor in these window regions could explain many of the observed discrepancies. In fact, Rawlins and Foot (1990), Nakajima et al. (1991), and, we suspect, Twomey and Cocks (1989), all used LOWTRAN 5 (Kneizys et al., 1980) to provide the water vapor absorption coefficients used in their retrieval schemes. LOWTRAN 7 (Kneizys et al., 1988) differs substantially from LOWTRAN 5 in its water vapor continuum and absorption line parameters, as well as in its pressure and temperature scaling for inhomogeneous vertical paths. Recently Taylor (1992) reanalyzed many of the data sets previously reported by Rawlins and Foot (1990), this time using LOWTRAN 7. He found that the bias in the retrieved values of r_e that Rawlins and Foot (1990) found when they used LOWTRAN 5 largely disappeared when the water vapor transmission characteristics of LOWTRAN 7 were used. In addition, Nakajima et al. (1991) demonstrated that their biases in r_e were the most consistent with excess absorption by water vapor in the near-infrared and were completely inconsistent with excess absorption within the cloud droplets or interstitial to the droplets.

In spite of the discrepancy that still remains between *in situ* and remote sensing estimates of r_e, it is nevertheless intriguing to examine the joint probability density function of τ_c and r_e for marine stratocumulus clouds, especially given the recent interest in parameterizing the shortwave radiative properties of clouds in terms of liquid water path (W) and effective radius (Slingo, 1989). Figure 9a shows joint probability density functions of τ_c and r_e derived from MCR images on each of four days during FIRE (after Nakajima et al., 1991), where each probability density function was derived from an image 35 km in width and 105–165 km in length, depending on the day. The contour lines for each day correspond to the 10, 30, 50, 70, and 90% occurrence levels, from which the mode and interquartile ranges can readily be inferred. Comparable results for W and r_e are presented in Fig. 9b, where the cloud optical thickness was converted to liquid water path using the relation (Stephens, 1978)

$$W = \frac{2\rho}{3} \tau_c r_e \tag{5}$$

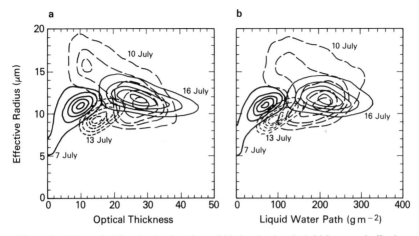

Figure 9 Joint probability density functions of (a) the cloud optical thickness and effective radius, and (b) the liquid water path and effective radius, for four days during FIRE. These results were derived from MCR measurements acquired from the ER-2 aircraft, where the effective radius r_{remote} has not been adjusted to the geometric center of the cloud layer. The five contour lines for each day correspond to the 10, 30, 50, 70, and 90% occurrence levels. (From Nakajima et al., 1991.)

where W is the liquid water path (g m^{-2}), ρ is the density of water (g cm^{-3}), and r_e is the effective radius (μm).

Figure 9 shows a distinct positive correlation between τ_c (or W) and r_e on the optically thin days of 7 and 13 July, and a modest negative correlation on the optically thick days of 10 and 16 July. Statistical properties like those presented in Fig. 9 are extremely important for climate studies, not simply because the joint retrieval of τ_c and r_e seems possible, but because the shortwave radiative properties of water clouds depend almost exclusively on these two parameters (Slingo, 1989).

C. Variability of Spectral Reflectance

On two different days during the FIRE marine stratocumulus experiment, the University of Washington C-131A and United Kingdom C-130 aircraft flew tightly coordinated flight tracks above, within, and below clouds. Figure 10 shows an intercomparison of the reflected intensity obtained on one such occasion (5 July 1987) when both aircraft were flying wingtip to wingtip for calibration intercomparison purposes. Each aircraft observed the reflected intensity in the nadir direction using its respective narrow field-of-view radiometer: the scanning cloud absorption radiometer with its 13 visible and near-infrared channels in the case of the C-131A (King et al., 1986), and the nadir-viewing multichannel radiometer with its eight visible and near-infrared channels in the case of the C-130 (Foot,

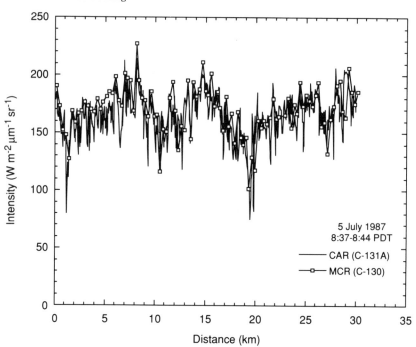

Figure 10   Comparison of the reflected intensity as a function of distance along the flight track of University of Washington C-131A and United Kingdom C-130 aircraft. These data were obtained above marine stratocumulus clouds on 5 July 1987 using the cloud absorption radiometer at 0.5 μm (solid line) and the multichannel radiometer at 0.55 μm (open squares).

1988; Rawlins and Foot, 1990). Figure 10 clearly shows that the visible calibration was quite similar for both radiometers on this day, and further that the reflected intensity of these marine stratocumulus clouds exhibited a cellular spatial structure. On this day the reflectance of the clouds varied with a wavelength of approximately 10 km, even though the clouds were geometrically quite flat on top. We attribute these reflectance variations to variations in the optical thickness (liquid water path) and, to a lesser extent, effective radius. We would further like to emphasize that these clouds, although generally flat on top, often exhibited variations in the cloud base, necessarily leading to variations in the geometric thickness of the clouds.

D.   Angular Transmission Characteristics

In addition to angular and spectral reflectance measurements and internal scattered radiation measurements, information on the radiative properties of clouds can be determined from the angular distribution of transmitted radiation beneath an optically thick cloud layer. This was shown by Rawlins and Foot (1990), who utilized zenith measurements of transmitted radiation beneath clouds to determine τ_c and r_e. They found that transmission measurements, although far less sensitive to microphysical properties than reflectance measurements, yield estimates of the effective radius that are entirely consistent with near-infrared reflectance measurements. The self-consistency of their reflection and transmission-derived values of r_e negates the possibility that their retrieval bias, discussed in Section II.B, could have arisen from a calibration error. This is because any calibration adjustment required to bring the spectral reflectance estimates into agreement with *in situ* measurements would necessarily lead to accentuating the bias errors resulting from the transmission measurements.

A further example of the use of transmission measurements can be found in Fig. 11, which shows the intensity as a function of zenith angle for $\lambda = 0.673$ μm and a single scan of the cloud absorption radiometer beneath a marine stratocumulus cloud layer on 13 July 1987. The angular distribution of the transmitted radiation beneath the cloud layer is seen to monotonically decrease from zenith to horizon, in close accord with our expectations for optically thick cloud layers in which the transmitted intensity $I(\tau_c; \mu, \phi)$ for conservative scattering is expected to be azimuthally independent and to follow the functional form (King, 1987)

$$I(\tau_c; \mu, \phi) = \frac{4\mu_0 F_0 K(\mu_0)[(1 - A_g)K(\mu) + A_g]}{\pi[3(1 - A_g)(1 - g)(\tau_c + 2q_0) + 4A_g]} \qquad (6)$$

In this expression q_0 is the extrapolation length, representing a virtual optical depth beneath the cloud layer from which the transmitted solar radiation is reflected back up to the cloud, $q' = (1 - g)q_0$ is the reduced extrapolation length ($\simeq 0.714$), $K(\mu)$ is the escape function, and all other constants have previously been defined. From this expression, it follows that the transmitted intensity beneath an optically thick cloud layer ($\tau_c \gg 2q_0 \simeq 9.4$) overlying a low-reflectance surface (such as the ocean) is given approximately by

$$I(\tau_c; \mu, \phi) \propto \frac{K(\mu)}{(1 - g)\tau_c} \qquad (7)$$

Thus, it follows that an angular transmission measurement, such as the one shown in Fig. 11, in conjunction with a theoretical calculation of the escape function (King, 1987), can easily be used to determine the optical thickness of a cloud layer. For an optically thick cloud, such as the one from which the measurements in Fig. 11 were obtained, the transmitted intensity is inversely proportional to the scaled optical thickness $(1 - g)\tau_c$. The low reflectance for $\theta \geq 90°$ corresponds to reflectance by the ocean surface.

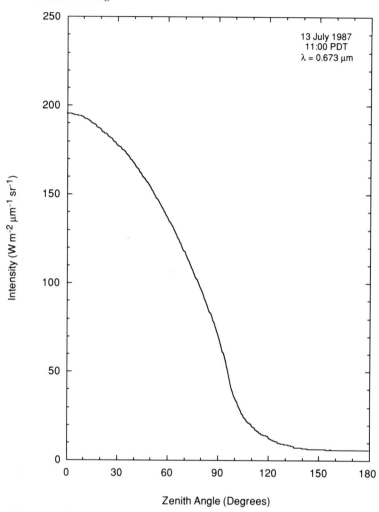

Figure 11   Intensity as a function of zenith angle for measurements obtained with the cloud absorption radiometer beneath clouds at λ = 0.673 μm. This scan applies to measurements taken over the ocean on 13 July 1987.

III.   Cirrus Clouds

Cirrus clouds were the focus of intensive field observations in the United States in 1986 (Starr, 1987) and 1991 (FIRE II cirrus IFO), and in Europe in 1989 (International Cirrus Experiment). These clouds, like marine stratocumulus clouds, are sensitive regulators of the earth's climate (Ramanathan et al., 1983, 1989) and are difficult to sense from remote satellite platforms. In the case of optically thin cirrus clouds, they often have little effect on reflected solar radiation but have a significant influence on the infrared radiative properties of the earth–atmosphere–ocean system. In the following sections we will describe some of the key findings on the radiative properties of cirrus clouds deduced from aircraft observations during the FIRE cirrus IFO, conducted in south-central Wisconsin during October and November 1986.

A.   Thermal Emission Characteristics

Figure 12 shows measurements of the brightness temperature spectrum observed between 600 and 1100 cm^{-1} (9.1 and 16.7 µm) using the nadir-viewing high-resolution interferometer sounder (HIS) flown on the ER-2 aircraft during 2 November 1986. Major absorption bands in the earth's atmosphere are evident in this figure as lower temperatures, representing emission from layers of the atmosphere up to 70 K colder than the earth's surface. The dominant absorption bands in this figure are the 15-µm CO_2 band (580 $\lesssim \nu \lesssim$ 760 cm^{-1}) and the 9.6-µm O_3 band (1000 $\lesssim \nu \lesssim$ 1070 cm^{-1}). The four curves presented in this figure represent measurements of upwelling (zenith propagating) thermal radiation from clear scenes

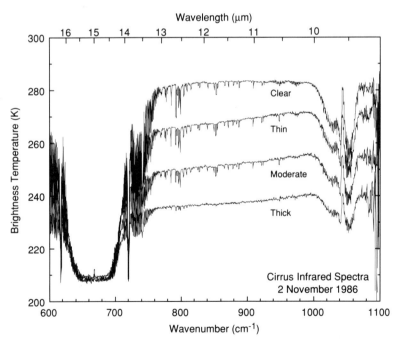

Figure 12   HIS brightness temperature spectra obtained between 600 and 1100 cm^{-1} (9.1 and 16.7 µm) over clear scenes and optically thin, moderate, and thick cirrus clouds during the FIRE cirrus experiment. (From King et al., 1992.)

and optically thin, moderate, and thick cirrus clouds. Over the clear land scene, the thermal emission in the 11-μm window region is spectrally quite flat, suggesting that the emissivity of the land surface is nearly the same at 11 and 12 μm (i.e., $\epsilon_{11} \simeq \epsilon_{12} \simeq 1.0$). In contrast, cirrus clouds tend to emit radiation with a higher brightness temperature at 11 μm than at 12 μm. This is because the emissivity of these optically thin ice clouds varies with wavelength and is everywhere less than unity ($\epsilon_{12} \geq \epsilon_{11}$). As the optical thickness increases and the emissivity of the clouds approaches unity, the brightness temperature difference once again disappears, as in the case of clear skies. In this case, however, the temperature of the scene is colder, corresponding to the temperature at the tops of the clouds. This principle, discussed in detail by Wu (1987) and Prabhakara et al. (1988), necessarily leads to the "droop" in the brightness temperature ($T_{11} - T_{12}$) observed in these measurements. Further examples of infrared emission spectra of cirrus clouds obtained from the Nimbus 4 interferometer spectrometer (IRIS) can be found in Prabhakara et al. (1990).

B.   Effective Radius of Ice Crystals

Prabhakara et al. (1988) were among the first to demonstrate the sensitivity of the brightness temperature difference ($T_{11} - T_{12}$) to effective particle radius and infrared optical thickness of cirrus clouds. Figure 13 shows calculations of the brightness temperature difference ($T_{10.8} - T_{12.6}$) as a function of the brightness temperature at 10.8 μm ($T_{10.8}$) for various values of the effective radius and infrared optical thickness. These calculations, based on a tropical atmosphere with 4 g cm^{-2} of precipitable water in which there are spherical ice crystals in a 1 km thick cloud near 9 km altitude (240 K) overlying an ocean with a sea-surface temperature of 300 K, show that the brightness temperature difference is everywhere positive. Furthermore, the temperature difference tends to be small for clear skies ($\tau_c = 0$, $T_{10.8} = 293$ K) and for optically thick clouds ($\tau_c \gtrsim 4$, $T_{10.8} \simeq 240$ K), with the greatest brightness temperature difference associated with small ice crystals and $\tau_c \simeq 1$, where the optical thickness here refers to the optical thickness at 10.8 μm.

Prabhakara et al. (1988) determined the global distribution of this brightness temperature difference from Nimbus 4 IRIS observations, and concluded that seasonal mean differences in excess of 8 K were frequently encountered in the tropical oceans, especially in the warm pool region of the western Pacific. They attribute these differences to the high frequency of optically thin cirrus clouds composed of small ice crystals. In addition to these satellite observations, Ackerman et al. (1990) examined the infrared radiative properties of cirrus clouds using HIS observations from the ER-2 aircraft during the FIRE cirrus IFO. They noted that the 8.55 μm window region is centered on a region containing weak water-vapor absorption lines, whereas 11 and 12 μm are in windows largely unaffected by water-vapor absorption lines. They further showed that the brightness temper-

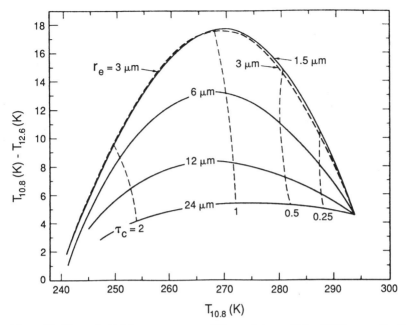

Figure 13 Theoretical brightness temperature difference between 10.8 and 12.6 μm as a function of the corresponding brightness temperature at 10.8 μm for high-level (cirrus) spherical ice crystal clouds as a function of effective radius and infrared optical thickness. (From Prabhakara et al., 1988.)

ature at 8.55 μm ($T_{8.5}$) generally exceeds that at 11 μm (T_{11}) over optically thin cirrus clouds, but it is less over clear sky regions. As a consequence, the brightness temperature difference $T_{8.5} - T_{11}$ is an even more sensitive indicator of optically thin cirrus clouds than is the brightness temperature difference $T_{11} - T_{12}$. Based on their observations during FIRE, they concluded that cirrus clouds are often composed of small ice crystals, with 8% of the cases having $10 \leq r_e < 30$ μm, 80% of the cases having $30 \leq r_e \leq 40$ μm, and 12% of the cases having $r_e > 40$ μm.

In addition to thermal emission measurements obtained with the nadir-viewing HIS, the ER-2 aircraft flown during the FIRE cirrus IFOs contained imaging radiometers and a monostatic Nd:YAG cloud and aerosol lidar system (Spinhirne et al., 1982, 1983). Plate 1, from Spinhirne and Hart (1990), shows a cross section of the lidar depolarization ratio as a function of distance as the aircraft flew over multilayer clouds on 28 October 1986, where the depolarization ratio δ is here defined as the ratio of the lidar return signal polarized perpendicular ($P_{\perp}$) to that polarized parallel ($P_{\parallel}$) to the transmitted laser pulse:

$$\delta = P_{\perp}/P_{\parallel} \tag{8}$$

The pulse repetition rate of the 0.532-μm laser was 5 Hz, resulting in a horizontal sampling interval of 40 m at the nominal aircraft speed of 200 m s^{-1}. This figure clearly shows that in the optically thin upper (cirrus) cloud the depolarization ratio typically ranged between 0.4 and 0.5, strongly suggesting the presence of nonspherical ice crystals. The lower (altocumulus) cloud layer, on the other hand, exhibited low values of the depolarization ratio ($\lesssim$ 0.05 near cloud top at an altitude of ~7.4 km), indicative of backscattering from spherical water droplets. The presence of supercooled water droplets in the altocumulus clouds and nonspherical ice crystals in the cirrus clouds was confirmed from nearly coincident *in situ* microphysical measurements by Heymsfield et al. (1990). The vertical cloud streamer located between 7.2 and 7.5 km at a distance of ~180 km is apparently due to an aircraft contrail, perhaps that of the *in situ* King Air aircraft.

From scanning radiometer measurements on the ER-2 aircraft, Spinhirne and Hart (1990) determined the brightness temperature at 11.2 μm ($T_{11.2}$) and the brightness temperature difference $\Delta T = T_{11.2} - T_{12.4}$ as a function of flight track distance. These results, presented in Fig. 14, correspond to the same time interval and aircraft flight track as presented in Plate 1. These results show that when the scene beneath the aircraft consists exclusively of optically thin cirrus clouds, as between 30 and 100 km, the brightness temperature difference lies in the range

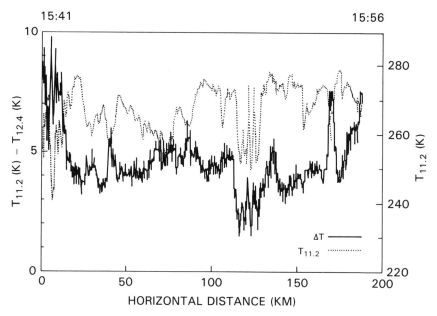

Figure 14   Brightness temperature at 11.2 μm and the brightness temperature difference $\Delta T = T_{11.2} - T_{12.4}$ as a function of flight track distance for nadir observations of cirrus clouds by the ER-2 on 28 October 1986. The brightness temperature difference varies with distance primarily as a result of the presence and optical characteristics of the lower cloud layer. (From Spinhirne and Hart, 1990.)

$4 \leqslant \Delta T \leqslant 5$ K. Calculations, such as the ones presented by Prabhakara et al. (1988) and shown in Fig. 13, therefore suggest that the upper-level cirrus cloud was composed of ice crystals having an effective radius in the range $12 \leqslant r_e \leqslant 24$ μm. In the portion of the scene containing cirrus clouds above optically thick altocumulus clouds, the brightness temperature difference ΔT is even larger than in the case of cirrus clouds alone. This is because these altocumulus clouds, found near the beginning and end of the flight line, are composed primarily of small water droplets. Had the lidar not been available to distinguish the multilayer nature of this cloud system, the infrared data alone would likely have been misinterpreted as suggesting the presence of quite small ice crystals. In the region between 120 and 130 km, where a lower-level cirrus cloud occurs in the absence of either an upper-level cirrus or lower-level altocumulus cloud, the brightness temperature difference is once again small, suggesting a region having relatively large ice crystals. In general, the cirrus cloud layer above 9 km was composed of the smallest ice crystals, those having $r_e \leqslant 25$ μm, and the lower-level cirrus cloud layer to ice crystals for which $r_e > 25$ μm.

Finally, Wielicki et al. (1990) used near-infrared reflectance measurements to estimate the effective radius of ice crystals in cirrus clouds using Landsat thematic mapper (TM) data obtained during the FIRE cirrus IFO. In general, they found that the satellite near-infrared reflection function measurements were largely consistent with values of $r_e \simeq 60$ μm, again far smaller than available *in situ* microphysical measurements ($r_e \simeq 200$ μm). All of these radiation measurements, taken together, suggest that the effective particle radius of ice crystals in the earth's atmosphere is much smaller than our current ability to measure it. This difficulty arises primarily from the fact that the principal cloud microphysics probes used today are based on light-scattering properties of spherical particles. Much effort needs to be expended in improving *in situ* particle-sizing capability, especially for the small ice particle sizes most often inferred from remote sensing of cirrus clouds.

C. Relationship between Thermal Emittance and Visible Albedo

The relationship between the thermal emission and shortwave reflection properties of cirrus clouds is important for assessing the radiative impact of cirrus clouds on the earth's climate (Arking, 1991). During the FIRE cirrus IFO Spinhirne and Hart (1990) measured the thermal emission at 10.84 μm and the reflection function at 0.75 μm using the narrow field of view MCR. Nadir observations of these parameters were obtained over cirrus clouds and clear sky conditions for the entire ER-2 flight track on 28 October 1986, a flight track that extended from central Wisconsin to overflights of Lake Michigan. After correcting this extensive data set for variable surface reflectance and further converting the nadir reflection function to plane albedo in accordance with the bidirectional reflectance model of Platt et al. (1980), Spinhirne and Hart (1990) were able to determine the relation-

146   Michael D. King

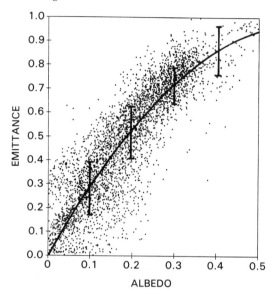

Figure 15   The effective emittance at 10.8 μm as a function of plane albedo at 0.75 μm for all measurements along the nadir track of the ER-2 on 28 October 1986. The curve represents a third-order polynomial fit to the observational data, and the error bars represent two standard deviations of the emittance averaged over 0.01 albedo intervals. (From Spinhirne and Hart, 1990.)

ship between nadir beam emittance and plane albedo. These results, presented in Fig. 15, were fitted to a third-order polynomial resulting in an expression of the following form:

$$\epsilon^\uparrow_{10.8} = 0.1456a^3_{0.75} - 2.677a^2_{0.75} + 3.185a_{0.75} \tag{9}$$

In this expression, strictly applicable for $0.0 \leqslant a_{0.75} \leqslant 0.45$, $\epsilon^\uparrow_{10.8}$ denotes the zenith-propagating beam emittance at 10.8 μm, and $a_{0.75}$ is the plane albedo at 0.75 μm. In addition to this parameterization, Spinhirne and Hart (1990) computed the standard deviation of the thermal beam emittance averaged over 0.01 albedo intervals, represented in Fig. 15 as error bars of one standard deviation on either side of the regression curve.

The relationship between visible reflectance and thermal emittance is also a major factor affecting the determination of cloud-top altitude and temperature from satellite observations. The thermal emittance determined by Spinhirne and Hart (1990) is decidedly less than that assumed by ISCCP for the global processing of satellite data at a given value of the visible reflectance (Rossow and Schiffer, 1991). This undoubtedly contributes to biases in the derivation of cloud-top altitude and temperature using ISCCP or NOAA-5 Scanning Radiometer (SR) algorithms (Rossow et al., 1989; Rossow and Schiffer, 1991), and points to the dire need to obtain further simultaneous observations of the spectral reflection and thermal emission properties of real clouds in a real atmosphere.

IV. Concluding Remarks

On examining the state of our knowledge of the radiative properties of clouds, we find that there has been a longstanding tendency for too many theories to chase too few observations. One of the most encouraging aspects of work carried out in recent years, which has been discussed in this review, is that increasing attention is being focused on new and innovative observational approaches and measurement techniques. These new approaches and instrumentation concepts make extensive use of both spectral and angular information, thus permitting greater examination of discrepancies between theory and observations than do traditional broadband flux observations. The growing use of cloud microphysics measurements to verify our interpretation of spectral cloud radiation observations is to be encouraged in the future.

Finally, it is important to recognize that much of the analysis that has thus far been applied to the remote sensing of cloud properties from aircraft and space-borne platforms has been based on applications of Mie theory for light scattering by spherical particles, and plane-parallel radiative transfer theory. There is a need to extend and revise these simple models, which are often able to explain the majority of the observed radiative properties of clouds, to problems associated with horizontal inhomogeneities in real clouds. These complexities have as yet not been fully explored using both theoretical calculations and experimental observations. This may very well limit the utility of many of the retrieval schemes discussed in this chapter to the special, simple cases to which they have thus far been applied.

References

Ackerman, S. A., and Cox, S. K. (1981). *J. Appl. Meteor.* **20**, 1510–1515.
Ackerman, S. A., Smith, W. L., Spinhirne, J. D., and Revercomb, H. E. (1990). *Mon. Wea. Rev.* **118**, 2377–2388.
Arking, A. (1991). *Bull. Amer. Meteor. Soc.* **71**, 795–813.
Arking, A., and Childs, J. D. (1985). *J. Climate Appl. Meteor.* **24**, 322–333.
Cess, R. D., Potter, G. L., Blanchet, J. P., Boer, G. J., Ghan, S. J., Kiehl, J. T., Le Treut, H., Li, Z. X., Liang, X. Z., Mitchell, J. F. B., Morcrette, J. J., Randall, D. A., Riches, M. R., Roeckner, E., Schlese, U., Slingo, A., Taylor, K. E., Washington, W. M., Wetherald, R. T., and Yagai, I. (1989). *Science* **245**, 513–516.
Cox, S. K., McDougal, D. S., Randall D. A., and Schiffer, R. A. (1987). *Bull. Amer. Meteor. Soc.* **68**, 114–118.
Curran, R. J., and Wu, M. L. C. (1982). *J. Atmos. Sci.* **39**, 635–647.
Curran, R. J., Kyle, H. L., Blaine, L. R., Smith, J., and Clem, T. D. (1981). *Rev. Sci. Instrum.* **52**, 1546–1555.
Davies, R., Ridgway, W. L., and Kim, K. E. (1984). *J. Atmos. Sci.* **41**, 2126–2137.
Foot, J. S., (1988). *Quart. J. Roy. Meteor. Soc.* **114**, 129–144.
Hansen, J. E., and Pollack, J. B. (1970). *J. Atmos. Sci.* **27**, 265–281.
Hansen, J. E., and Travis, L. D. (1974). *Space Sci. Rev.* **16**, 527–610.

Henyey, L. C., and Greenstein, L. J. (1941). *Astrophys. J.* **93**, 70–83.

Herman, B. M., Asous, W., and Browning, S. R. (1980). *J. Atmos. Sci.* **37**, 1828–1838.

Herman, G. F. (1977). *J. Atmos. Sci.* **34**, 1423–1432.

Herman, G. F., and Curry, J. A. (1984). *J. Climate Appl. Meteor.* **23**, 5–24.

Heymsfield, A. J., Miller, K. M., and Spinhirne, J. D. (1990). *Mon. Wea. Rev.* **118**, 2313–2328.

Hignett, P. (1987). *Quart. J. Roy. Meteor. Soc.* **113**, 1011–1024.

King, M. D. (1981). *J. Atmos. Sci.* **38**, 2031–2044.

King, M. D. (1987). *J. Atmos. Sci.* **44**, 1734–1751.

King, M. D., Strange, M. G., Leone, P., and Blaine, L. R. (1986). *J. Atmos. Oceanic Tech.* **3**, 513–522.

King, M. D., Radke, L. F., and Hobbs, P. V. (1990). *J. Atmos. Sci.* **47**, 894–907.

King, M. D., Kaufman, Y. J., Menzel, W. P., and Tanré, D. (1992). *IEEE. Trans. Geosci. Remote Sens.* **29**, 2–27.

Kneizys, K. X., Shettle, E. P., Gallery, W. O., Chetwynd, J. H., Abreu, L. W., Selby, J. E. A., Fenn, R. W. and McClatchey, R. A. (1980). Atmospheric transmittance/radiance: Computer code LOWTRAN 5. AFGL-TR-80-0067, Air Force Geophysics Laboratories, Hanscom AFB, 233 pp.

Kneizys, K. X., Shettle, E. P., Abreu, L. W., Chetwynd, J. H., Anderson, G. P., Gallery, W. O., Selby, J. E. A., and Clough, S. A. (1988). Users guide to LOWTRAN 7. AFGL-TR-88-0177, Air Force Geophysics Laboratories, Hanscom AFB, 137 pp.

Mel'nikova, I. N. (1978). *Izv. Acad. Sci., USSR, Atmos. Ocean. Phys.* **14**, 928–931.

Nakajima, T., and King, M. D. (1990). *J. Atmos. Sci.* **47**, 1878–1893.

Nakajima, T., King, M. D., Spinhirne, J. D., and Radke, L. F. (1991). *J. Atmos. Sci.* **48**, 728–750.

Newiger, M., and Bähnke, K. (1981). *Contr. Atmos. Phys.* **54**, 370–382.

Platt, C. M. R., Reynolds, D. W., and Abshire, N. L. (1980). *Mon. Wea. Rev.* **108**, 195–204.

Prabhakara, C., Fraser, R. S., Dalu, G., Wu, M. L. C., and Curran, R. J. (1988). *J. Appl. Meteor.* **27**, 379–399.

Prabhakara, C., Yoo, J. M., Dalu, G., and Fraser, R. S. (1990). *J. Appl. Meteor.* **29**, 1313–1329.

Ramanathan, V. (1987). *J. Geophys. Res.* **92**, 4075–4095.

Ramanathan, V., Pitcher, E. J., Malone, R. C., and Blackmon, M. L. (1983). *J. Atmos. Sci.* **40**, 605–630.

Ramanathan, V., Cess, R. D., Harrison, E. F., Minnis, P., Barkstrom, B. R., Ahmad, E., and Hartmann, D. (1989). *Science* **243**, 57–63.

Rawlins, F., and Foot, J. S. (1990). *J. Atmos. Sci.* **47**, 2488–2503.

Reynolds, D. W., Vonder Haar, T. H., and Cox, S. K. (1975). *J. Appl. Meteor.* **14**, 433–444.

Rossow, W. B., and Schiffer, R. A. (1991). *Bull. Amer. Meteor. Soc.* **72**, 2–20.

Rossow, W. B., Gardner, L. C., and Lacis, A. A. (1989). *J. Climate* **2**, 419–458.

Schiffer, R. A., and Rossow, W. B. (1983). *Bull. Amer. Meteor. Soc.* **64**, 779–784.

Slingo, A. (1989). *J. Atmos. Sci.* **46**, 1419–1427.

Slingo, A., and Schrecker, H. M. (1982). *Quart. J. Roy. Meteor. Soc.* **108**, 407–426.

Spinhirne, J. D., and Hart, W. D. (1990). *Mon. Wea. Rev.* **118**, 2329–2343.

Spinhirne, J. D., Hansen, M. Z., and Caudill, L. O. (1982). *Appl. Opt.* **21**, 1564–1571.

Spinhirne, J. D., Hansen, M. Z., and Simpson, J. (1983). *J. Climate Appl. Meteor.* **22**, 1319–1331.

Starr, D. O'C. (1987). *Bull. Amer. Meteor. Soc.* **68**, 119–124.

Stephens, G. L. (1978). *J. Atmos. Sci.* **35**, 2123–2132.

Stephens, G. L., and Platt, C. M. R. (1987). *J. Climate Appl. Meteor.* **26**, 1243–1269.

Stephens, G. L., and Tsay, S. C. (1990). *Quart. J. Roy. Meteor. Soc* **116**, 671–704.

Stephens, G. L., Paltridge, G. W., and Platt, C. M. R. (1978). *J. Atmos. Sci.* **35**, 2133–2141.

Stephens, G. L., Ackerman, S., and Smith, E. A. (1984). *J. Atmos. Sci.* **41**, 687–690.

Stowe, L. L., Yeh, H. Y. M., Eck, T. F., Wellemeyer, C. G., Kyle, H. L., and the Nimbus-7 Cloud Data Processing Team (1989). *J. Climate* **2**, 671–709.

Taylor, J. P. (1992). *J. Atmos. Sci.* **49**, 2564–2569.

Twomey, S. (1972). *J. Atmos. Sci.* **29**, 1156–1159.

Twomey, S. (1976). *J. Atmos. Sci.* **33**, 1087–1091.
Twomey, S. (1977). *J. Atmos. Sci.* **34**, 1149–1152.
Twomey, S. (1983). *Contrib. Atmos. Phys.* **56**, 429–439.
Twomey, S., and Cocks, T. (1982). *J. Meteor. Soc. Japan* **60**, 583–592.
Twomey, S., and Cocks, T. (1989). *Beitr. Phys. Atmos.* **62**, 172–179.
Welch, R. M., Cox, S. K., and Davis, J. M. (1980). *Solar Radiation and Clouds.* Meteor. Monogr. No. 39, *Amer. Meteor. Soc.,* 96 pp.
Wielicki, B. A., Suttles, J. T., Heymsfield, A. J., Welch, R. M., Spinhirne, J. D., Wu, M. L. C., Starr, D. O'C., Parker, L., and Arduini, R. F. (1990). *Mon. Wea. Rev.* **118**, 2356–2376.
Wiscombe, W. J., Welch, R. M., and Hall, W. D. (1984). *J. Atmos. Sci.* **41**, 1336–1355.
Wu, M. L. C. (1987). *J. Appl. Meteor.* **26**, 225–233.

Chapter 6 | Radiative Effects of Clouds on Earth's Climate

D. L. Hartmann
Department of Atmospheric Sciences
University of Washington
Seattle, Washington

The radiative effects of clouds on the energy balance of Earth have been studied for about a century. In recent years observations from earth-orbiting satellites have greatly improved the quality and detail of the information available about the global distribution of clouds and their effects on the energy balance of Earth. Data from the Earth Radiation Budget Experiment provide a two-year data set based on observations taken from two identical scanning instruments flying simultaneously on two satellites in very different orbits. These data describe the diurnal, seasonal, and regional variations of cloud radiative forcing of the climate system.

Recent efforts have used radiation budget data in conjunction with cloud descriptions based on satellite data to estimate the importance of various cloud types in the energy balance of Earth. Low stratus clouds and tropical convective anvil clouds play very important roles in radiative forcing of climate. High-latitude clouds, especially those over the oceans, provide a strong negative cloud forcing during the summer, which results in an increase in the required equator-to-pole energy transport.

Cloud radiative effects play a potentially important role in determining the magnitude and geographical distribution of climate changes that result from natural or human forcing of climate change. Several possible climate feedbacks that involve cloud radiative forcing have been proposed. To test these hypotheses requires their incorporation in global climate models and the provision of adequate data to validate simulations of cloud processes. A combination of new and more detailed satellite measurements and aircraft data is needed to advance our understanding of cloud formation and radiative feedbacks.

I. Introduction

Suspensions of liquid water and ice in the atmosphere have a dramatic influence on the reflection and absorption of solar radiation and on the emission and absorp-

Aerosol–Cloud–Climate Interactions   **151**

tion of terrestrial radiation. These radiative effects of clouds are important for the energy balance of Earth, and for physical, dynamical, chemical, and biological processes within the climate system. This review is limited primarily to the large-scale effects of clouds on the radiative energy balance at the top of the atmosphere and at the surface, and the roles these effects may play in determining the sensitivity of global climate to external forcing. The optical properties of clouds are discussed in Chapter 5 of this volume (see also Fouquart et al., 1990). To present a complete account of the importance of the radiative effects of clouds for climate one would also need to discuss the important interactions of cloud radiative effects with microphysical and dynamical processes that determine the amount and optical properties of clouds, but these interactions will not be considered here.

Implicit in much of the discussion of clouds is the assumption that the radiative effects of liquid water and ice can logically be separated from those of water vapor. From the perspective of remote sensing and radiative transfer it makes sense to consider the liquid and solid forms of water separately, since their optical properties are so different from those of the vapor phase. From the perspective of cloud dynamics and thermodynamics, however, the three phases of water are closely coupled. For example, the upward motions that generate condensation in convective clouds also carry moisture upward and may lead to an increase in the water vapor present in the upper troposphere. Upper tropospheric water vapor makes a very important contribution to the atmospheric greenhouse effect.

II. Early Work

The importance of clouds for weather and climate is apparent to the most casual observer. The effect of a cumulus cloud passing in front of the sun on a spring day can be detected by at least two of the five human senses. Sophisticated instrumentation is not necessary to conclude that the effects of clouds on radiation are important for climate. Therefore, clouds and their radiative effects have always been an important part of the science of climatology. Much early work was too limited by a lack of accurate observations to define the quantitative effects of clouds in global climate. Global observations were not available, so most observational work was local in nature and usually concerned with diurnal and seasonal time scales. Nonetheless, early workers knew that clouds were important for the global albedo of Earth and the global energy balance (e.g., Abbot and Fowle, 1908). Arrhenius (1896) included a fairly detailed treatment of cloud radiative effects in his calculations of the effect of carbon dioxide on climate. The effect of clouds on the diurnal range of temperature was recognized from an early date, and the effect of downward longwave radiation from clouds was measured early in this century (e.g., Ångström, 1928).

A. Observational Studies

The paper by Aldrich (1919) is an example from the classical literature that pre-figures recent interest in assessing the role of clouds in the global radiation balance. Aldrich reported on some broadband measurements of the albedo of valley clouds taken from a War Department observation balloon over Arcadia, California. On the morning the measurements were taken the fog was 500 meters thick in the early morning and thinned to about 180 meters by 11 AM. Aldrich and collaborators measured an albedo of 78% from this "cloud." Aldrich obtained a clear-sky albedo of 17% by assuming 8% is reflected from the surface and 9% from the atmosphere. It was believed that clouds covered 52% of Earth's surface (Bort, 1884), so that the measured cloud albedo of 78% would give a planetary albedo of 43%. This estimate was an increase from the earlier estimate of Abbot and Fowle (1908) and survived to be reported in Brunt's classic textbook (Brunt, 1934). Danjon (1936) estimated the albedo to be 0.39, based on visible measurements of the new moon. Fritz (1949) corrected Danjon's estimates for spectral biases and obtained a lower estimate of 0.35. Fritz was aware that the cloud albedo taken by Aldrich was probably high, based on measurements reported by Luckiesh (1919) and Neiburger (1949). Fritz inferred a global mean cloud albedo of about 0.5. A historical account of efforts to evaluate Earth's energy balance can be found in Hunt et al. (1986). Modern estimates place Earth's albedo near 0.3, its clear-sky albedo about 0.15, and fractional cloud cover about 0.6. These values imply an average cloudy-sky albedo of 0.4.

Haurwitz (1948) used regression to derive empirical relationships between surface insolation and cloud type, based on eight years of data from the Blue Hill Observatory. Only data for completely overcast skies were used. He concluded that high clouds reduced surface insolation about 20%, middle clouds 50 to 60%, and low clouds 65 to 80%. These numbers were employed by modelers and have persisted in the theoretical literature for more than 30 years. In the 1950s and 1960s, surface-based observations of cloud distributions were used in attempts to estimate the energy balance for Earth. Albedos in line with those suggested by Haurwitz (1948) were used for high (0.21), medium (0.48), and low (0.69) clouds (Houghton, 1954; London, 1956; Manabe and Strickler, 1964).

Measurements of Earth from space, which began in the late 1950s, have greatly improved estimates of the energy balance and the role of clouds in it (House et al., 1986). The Tiros satellite series and later the Nimbus satellite series produced steady improvement in our ability to measure clouds and radiation from space, and yielded reliable estimates of the geographical variations of albedo and outgoing longwave radiation (OLR) (Hartmann et al., 1986; Ohring and Gruber, 1983; Raschke et al., 1973; Stephens et al., 1981; Vonder Haar and Suomi, 1971).

B. Modeling

Manabe and Strickler (1964) presented a one-dimensional model of climate that would predict the global mean surface and air temperature from a thermodynamic balance. The optical properties and vertical distribution of clouds at three levels were specified and the resulting thermal equilibrium temperature profile was calculated by using a convective adjustment to represent the effects of nonradiative heat exchanges. The amount of cloud was specified from surface observations summarized by London (1956), Haurwitz's albedos were used, and the radiative effects of clouds at the surface and top of the atmosphere were broadly consistent with the results of London (1956) and Houghton (1954). The results illustrated the potential role of clouds in Earth's radiation balance and their effect on the global mean temperature in thermal equilibrium.

Manabe and Strickler obtained a global mean surface temperature very close to the observed value of 288 K when they included clouds in their one-dimensional model. Without clouds the computed temperature was about 13°C higher. The cloud cooling was accomplished primarily through a net decrease in absorbed solar radiation of about 70 W m^{-2}, which in their model corresponds to an albedo increase from 0.143 to 0.346 (Table 1). In an equilibrium calculation this reduction in absorbed solar radiation must be balanced by an equal reduction of the outgoing longwave radiation. The reduction in longwave energy loss when clouds are inserted is felt mostly in the atmosphere, where it is not compensated by a significant change in solar heating. Most of the top-of-atmosphere solar heating loss associated with the high albedo of clouds was transferred directly to the surface. It is partially offset there by a smaller decrease in the longwave energy lost from the surface. The net effect is a reduction of about 20% in the requirement for nonradiative surface cooling. Manabe and Wetherald (1967) repeated these calculations with a fixed relative humidity distribution, which increased the sensitivity of the model to all influences. Manabe and collaborators showed that low, bright clouds cool the surface temperature substantially, while high, thin clouds can warm the surface.

The earliest attempts to model the general circulation of the atmosphere did not explicitly take account of clouds or detailed radiative transfer, but used atmospheric heating rates calculated diagnostically (e.g., Smagorinsky, 1963). The first models to include radiative transfer followed the methodology of Manabe and Strickler (1964) by including uniform clouds with fixed optical properties (Smagorinsky et al., 1965). Later, clouds were specified as functions of latitude. Meleshko and Wetherald (1981) demonstrated the importance of longitudinal variations in cloud amount, especially in the Northern Hemisphere where land–sea differences are substantial.

Table 1

Effects of Cloudiness on the Energy Balance in the One-Dimensional Radiative/Convective Equilibrium Calculations of Manabe and Strickler (1964)[a]

	Clear	Cloudy	London	Forcing
OLR	299	228	226	−71
Absorbed solar	299	228	226	−71
Albedo	0.143	0.346		0.203
Atmospheric longwave heating	−195	−168	−163	27
Atmospheric shortwave heating	65	63	61	−2
Surface longwave heating	−104	−60	−63	44
Surface solar heating	234	165	165	−69
Nonradiative surface exchange	131	105	103	−25

[a]Units are W m^{-2}, except for albedo, which is a dimensionless fraction. Estimates of global means from London (1956) are shown for comparison. Forcing is difference between the cloudy and the clear calculation. (Used with permission from the American Meteorological Society.)

C. Cloud–Climate Sensitivity Research

Entering the 1970s, climate change became an important focus of research. The radiative effects of clouds quickly rose to prominence amid the myriad issues surrounding the question of climate change. A very useful discussion of the potential role of cloud radiative effects in climate sensitivity was presented by Schneider (1972). He emphasized the importance of cloud-top height and albedo for the energy balance and the general lack of data on these properties, especially for cirrus clouds. The 1970s and 1980s saw many efforts to use satellite observations to estimate the effect of clouds on the energy balance at the top of the atmosphere (Cess, 1976; Ellis, 1978; Hartmann and Short, 1980; Ohring et al., 1981; Ohring and Clapp, 1980; Cess et al., 1982). The problem of predicting climate change also forced climate modelers to face the necessity of predicting cloud radiative properties and their interaction with all other climate processes (Hansen et al., 1984; Hunt, 1982; Wetherald and Manabe, 1980).

III. Recent Advances

A. Observational Studies

The quality and detail of the observational data sets for cloud cover and Earth's radiation balance have greatly improved in the last decade as a result of the Nimbus-7 satellite, the Earth Radiation Budget Experiment (ERBE), and the International Satellite Cloud Climatology Project (ISCCP). The Nimbus-7 satellite included the ERB instrument (Jacobowitz et al., 1984; Kyle et al., 1985), which measured broadband energy fluxes, and the THIR instrument, which was used to provide a coincident cloud identification (Hwang et al., 1988; Stowe et al., 1989). The Earth Radiation Budget Experiment was the first multiple-satellite radiation budget experiment (Barkstrom and Smith, 1986), which included a medium-inclination orbiter to measure diurnal variations and used a scene-dependent inversion scheme (Smith et al., 1986; Wielicki and Green, 1989). ISCCP uses narrowband radiances in the visible and thermal infrared obtained from geostationary and polar-orbiting operational meteorological satellites to determine cloud coverage histograms binned according to cloud-top pressure and cloud visible optical depth (Rossow and Schiffer, 1991; Schiffer and Rossow, 1983).

1. Cloud Radiative Forcing

An important product of the ERBE experiment is the clear-sky climatology obtained by averaging together all scanner observations that are identified as cloud-free by the scene identification scheme. The difference between the cloud-free radiation budget climatology and the average over all scene types is the cloud radiative forcing, the amount by which the presence of clouds alters the top-of-atmosphere energy budget (Charlock and Ramanathan, 1985; Harrison et al., 1989; Ramanathan, 1987; Ramanathan et al., 1989). Cloud radiative forcing is a simple means by which to measure the effect of clouds on the radiation budget at the top of the atmosphere.

The net radiation at the top of the atmosphere is the difference between absorbed solar radiation and emitted terrestrial radiation:

$$R = S(1 - \alpha) - F = Q - F \qquad (1)$$

Here R is net radiation, S is insolation, α is planetary albedo, F is outgoing longwave radiation, and Q is absorbed solar radiation. We define the net cloud radiative forcing as the difference between the average and clear-sky radiative energy fluxes:

$$\text{Net cloud radiative forcing} = R - R_{\text{clear sky}} \qquad (2)$$

The shortwave cloud forcing is defined similarly:

$$\text{Shortwave cloud forcing} = Q - Q_{\text{clear sky}} \qquad (3)$$

The longwave cloud forcing is defined so that positive values indicate that cloud longwave effects increase the net radiation:

$$\text{Longwave cloud forcing} = F_{\text{clear sky}} - F \qquad (4)$$

Figures 1 through 4 show global distributions of cloud forcing for seasons based on two years when measurements were available from ERBE scanning instruments on both the ERBS and the NOAA-9 satellites (Feb. 1985–Jan. 1987). These data are accurate to ± 10 W m^{-2} in most regions, but are uncertain over snow-covered surfaces and in tropical regions where upper-tropospheric water-vapor variations are large (Hartmann and Doelling, 1991). Two-year averages of OLR, absorbed solar radiation, net radiation, and the cloud forcing of these quantities are shown in Plates 2, 3, and 4.

Longwave cloud forcing is defined here as the clear OLR minus the cloudy OLR, so that positive values indicate a positive effect of clouds on the energy budget. The values shown in Fig. 1 and Plate 2 are in broad agreement with other estimates based on METEOSAT observations (Schmetz et al., 1990). The longwave cloud forcing is largest where upper-level clouds are present, such as in regions of tropical convection and mid-latitude storm tracks. The largest values attained are about 80 W m^{-2}, and the global average value is about 30 W m^{-2}.

The effect of clouds on the albedo is also maximum in regions of active convection, but low clouds in stratus regions of the tropics and in high latitudes are of equal or greater importance (Fig. 2). The global average clear-sky value is about 15% which is half of the average value for all scenes. The effect of clouds on shortwave radiation is strongly modulated by the insolation, which is greatest in high latitudes during summer, where there are also substantial amounts of cloud and where the average solar zenith angle is large (Fig. 3 and Plate 3). Maximum

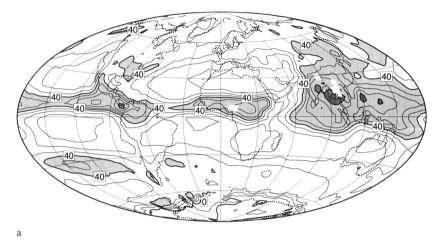

a

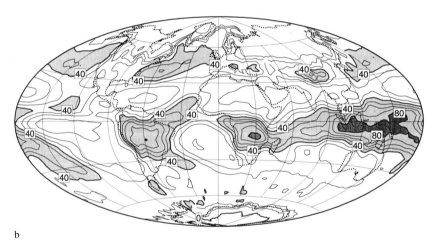

b

Figure 1 Longwave cloud radiative forcing of Earth's energy balance (a) for the June, July,
August (JJA) and (b) for December, January, February (DJF) seasons determined from two years
(Feb. 1985–Jan. 1987) of ERBE data from two scanning instruments on the ERBS and NOAA-9
satellites. The contour interval is 10 W m^{-2}. Values greater than $+40$ W m^{-2} are lightly shaded and
values greater than $+80$ W m^{-2} are heavily shaded. Note that positive values indicate that clouds
reduce the outgoing longwave radiation.

reductions in absorbed solar radiation exceed 120 W m^{-2} over the oceans in high
latitudes during summer and 100 W m^{-2} in tropical convective regions. Clouds
reduce the global average absorbed solar radiation by about 50 W m^{-2}.

The cloud forcing of net radiation at the top of the atmosphere is predominantly
negative and largest in magnitude over tropical stratus regions and over mid-

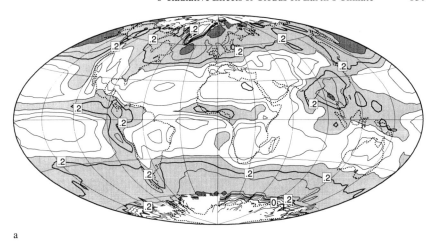

a

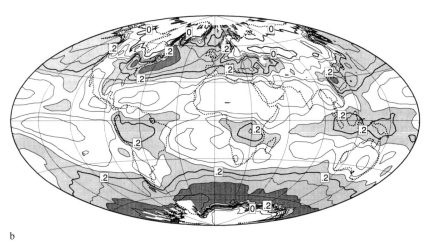

b

Figure 2 Cloud forcing of albedo (a) for JJA and (b) for DJF estimated from ERBE data. Contour interval is 0.05. Values greater than 0.15 are shaded and values greater than 0.3 are heavily shaded. Positive values indicate that clouds increase the albedo.

and high-latitude oceans during seasons with substantial insolation (Fig. 4 and Plate 4). The large longwave and shortwave cloud forcing in tropical convective regions largely cancel each other, giving a relatively small net forcing in these regions. The global average net cloud radiative forcing is about 20 W m^{-2}.

Zonal averages of net cloud radiative forcing show the importance of middle and high-latitude cloudiness, especially during the summer season when the insolation is large and the net cloud forcing can produce a cooling of 100 W m^{-2}

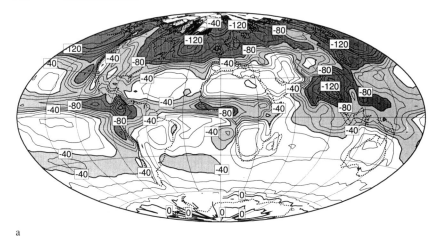

a

b

Figure 3   Cloud forcing of absorbed solar radiation (a) for JJA and (b) for DJF estimated from ERBE data. Contour interval is 20 W m^{-2}. Values more negative than -40 W m^{-2} are shaded and values more negative than -80 W m^{-2} are heavily shaded.

(Fig. 5c). Low, relatively abundant clouds and large average solar zenith angles give rise to large cloud-induced reductions in absorbed solar radiation that dominate the effects of clouds on net radiation in middle and high latitudes. Away from the polar regions longwave cloud forcing is about 30 W m^{-2}, but dips to 20 W m^{-2} in the subtropics and peaks near 50 W m^{-2} in the equatorial region. Between 35°N and 35°S the net cloud forcing is relatively small (about -15 W m^{-2}) and almost independent of latitude, especially for the annual average (Fig. 6). The small values and relative constancy are brought about by the

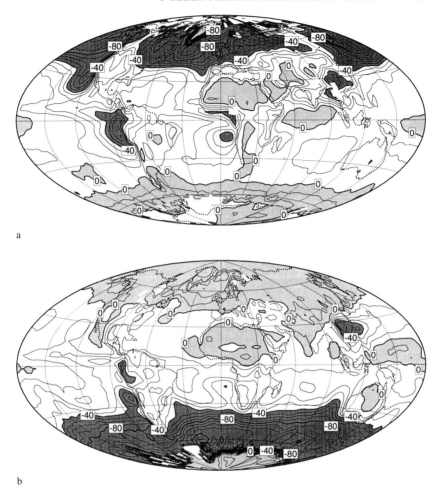

a

b

Figure 4   Net cloud radiative forcing of Earth's energy balance (a) for JJA and (b) for DJF estimated from ERBE data. Contour interval is 20 W m⁻². Positive values are shaded and values more
negative than − 40 W m⁻² are heavily shaded.

cancellation between solar and longwave contributions. The radiative effect of
clouds slightly increases the required equator-to-pole energy transport in the annual mean, since clouds reduce net radiation more in high latitudes than they do
in the tropics. This increased requirement for meridional heat transport is felt
mostly in the summer hemisphere, where the daily insolation is almost independent of latitude, but the reduction of net radiation by clouds in high latitudes main-
tains a substantial equatorward gradient in net radiation.

A summary of hemispheric and seasonal variations of the ERBE radiation bud-

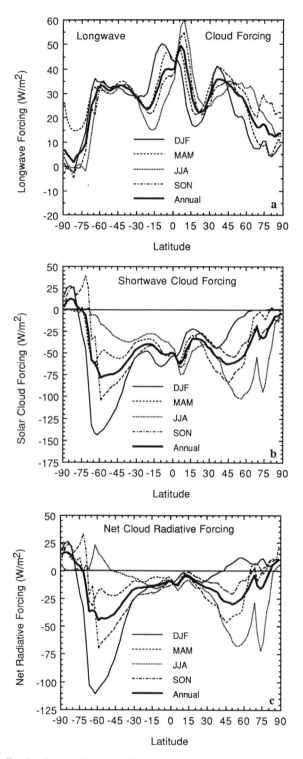

Figure 5 Zonal and seasonal averages of (a) longwave cloud forcing, (b) shortwave cloud forcing, and (c) net cloud radiative forcing based on two years of ERBE scanner data. Averages for each of four 3-month seasons and the annual mean are shown.

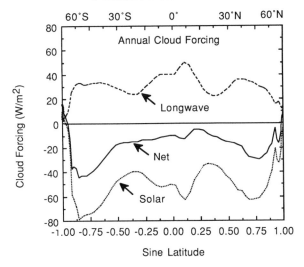

Figure 6   Zonal and annual averages of cloud forcing of longwave, solar, and net radiative fluxes into the climate system. These quantities are plotted as functions of the sine of latitude so that the area between the curve and the zero line is proportional to the contribution of each latitude to a global area average.

get climatology and cloud radiative forcing for a two-year climatology is given in Table 2. Because of the greater fraction of land, the Northern Hemisphere has a larger clear-sky albedo; but the Southern Hemisphere has a larger cloud albedo forcing, so that the average albedos of the two hemispheres are approximately the same. Cloud radiative effects reduce seasonal variations in the two hemispheres, because the cloud forcing reduces the amplitude of the annual cycle in hemispheric net radiation, primarily by reducing the absorption of solar radiation during summer. The cloud-caused reduction in the amplitude of the annual cycle in net radiation is about 28 and 36 W m^{-2} in the Northern and Southern Hemispheres, respectively.

2. Radiation Budget and Cloud Climatologies

Over the years considerable effort has been devoted to defining and measuring cloudiness, and it is reasonable to examine the relationships between these cloud climatologies and the net radiation at the top of the atmosphere and at the surface. A number of studies have looked at the relationship between satellite-derived estimates of the cloud cover and the radiation budget at the top of the atmosphere. Comparison of the Nimbus-7 radiation budget and cloudiness estimates yields estimates of cloud forcing that are similar to those derived from the ERBE clear-sky and average radiation budget climatologies (Ardanuy et al., 1991; Ardanuy et al., 1988). Dhuria and Kyle (1990) used Nimbus-7 ERB and cloud data to estimate the importance of clouds with different top temperatures estimated from the

Table 2

Average, Clear-Sky, and Cloud Forcing Estimates[a]

	Albedo			OLR			Absorbed solar			Net radiation		
	Ave.	Clr.	Fcg.	Ave.	Clr.	Fcg.	Ave.	Clr.	Fcg.	Ave.	Clr.	Fcg.
Northern Hemisphere												
DJF	.292	.172	.120	228	254	−26	168	196	−28	−59	−57	−2
MAM	.309	.169	.140	233	264	−31	269	320	−50	36	56	−20
JJA	.303	.149	.155	242	277	−35	307	371	−64	65	95	−30
SON	.289	.151	.138	234	267	−33	211	253	−42	−23	−14	−9
Annual	.298	.160	.138	234	265	−31	239	285	−46	5	20	−15
Southern Hemisphere												
DJF	.311	.142	.170	234	268	−34	320	395	−75	86	127	−41
MAM	.282	.134	.147	232	264	−32	208	252	−44	−25	−12	−12
JJA	.272	.143	.129	234	258	−24	162	190	−28	−72	−67	−5
SON	.303	.137	.166	234	261	−28	272	326	−54	38	65	−27
Annual	.292	.139	.153	233	263	−29	240	291	−50	7	28	−21
Global average												
DJF	.305	.152	.153	231	261	−30	244	296	−51	13	35	−21
MAM	.297	.154	.143	233	264	−31	239	286	−47	6	22	−16
JJA	.293	.147	.146	238	267	−29	234	281	−46	−3	14	−17
SON	.297	.143	.154	234	264	−30	241	289	−48	7	26	−18
Annual	.298	.149	.149	234	264	−30	240	288	−48	6	24	−18

[a] From two years of ERBE scanner observations (Feb. 1985–Jan. 1987) from the ERBS and NOAA-9 satellites.

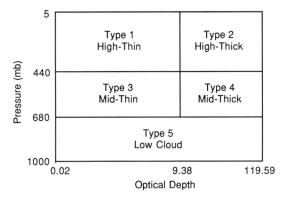

Figure 7 Diagram showing the division of the ISCCP cloud-top pressure and visible optical depth into five cloud types used in the regression.

estimated from the THIR instrument. Such investigations show the great importance of cloud type as well as amount. In many tropical convective regions the cloud cover remains almost uniformly high, while the radiation budget quantities vary widely in association with changes in the type of clouds present.

Hartmann et al. (1992) have taken simultaneous estimates of radiation budget quantities from ERBE and cloud distributions from ISCCP and related them via multiple regression. Comparison of these two data sets allows the estimation of the importance of particular cloud types for the energy budget at the top of the atmosphere. Radiation budget quantities are related to the fractional area coverage of clouds with visible optical depths and cloud-top within particular ranges (Fig. 7). The five cloud types used in this analysis are a great simplification of the 35 cloud types provided in the ISCCP C1 data set. Regression of radiation budget quantities on the fractional coverage by these five cloud types explains as much variance as any other grouping of the ISCCP C1 data. The regression equation was of the form:

$$R = R_0 + \sum_{i=1}^{5} \Delta R_i \, C_i \qquad (5)$$

where R is some radiation balance quantity; R_0 is a regression intercept, which is equated with the clear-sky value of the radiation balance quantity; C_i is the fractional coverage by one of five cloud types; and ΔR_i is the sensitivity coefficient relating changes in the radiation budget quantity to fractional coverage by clouds of type i. The results of this study are summarized in Table 3, which shows the global-average fractional area coverage by each cloud type and the cloud forcing by that type, as estimated by regression between daily means in $2.5° \times 2.5°$ regions of ERBE variables and ISCCP cloud fractions.

Table 3 indicates, as expected, that the highest clouds make the largest contri-

Table 3

Global Area-Averaged Cloud Forcing by Type of Cloud

Season: Parameters[a]	Type 1 high, thin		Type 2 high, thick		Type 3 mid, thin		Type 4 mid, thick		Type 5 low		Sum	
	JJA	DJF	JJA	DJF	JJA	DJF	JJA	DJF	JJA	DJF	Sum	Ave.
C_i	10.2	10.0	8.5	8.8	10.7	10.7	6.5	8.2	27.2	25.9		63.3
OLR	6.5	6.3	8.4	8.8	4.8	4.9	2.4	2.4	3.5	3.5		25.8
Albedo	1.2	1.1	4.1	4.2	1.1	1.0	2.7	3.0	5.8	5.6		14.9
Net	2.4	2.3	−6.4	−7.5	1.4	0.8	−6.6	−8.5	−15.1	−18.2		−27.6
ΔOLR	63.7	63.0	98.8	100.0	44.9	45.8	36.9	29.3	12.9	13.5		40.8
ΔAlbedo	11.8	11.0	48.2	47.7	10.3	9.3	41.5	36.6	21.3	21.6		23.5
ΔNet	23.5	23.0	−75.3	−85.2	13.1	7.5	−102.	−104.	−55.5	−70.3		−43.6

[a]OLR and net radiation are given in W m^{-2} and albedo and cloud fractional coverage (C_i) are given in percent. OLR, Albedo, and Net indicate the global-average forcing of the radiation balance by the cloud type of interest. ΔOLR, ΔAlbedo, and ΔNet indicate the average contrast between the cloud type and a clear scene and are obtained by dividing the global average cloud forcing by the global average cloud amount. The last column is the sum over all cloud types and the average of the JJA and DJF season. The regions poleward of about 60° in the winter hemisphere are not included in the area average.

bution to cloud longwave forcing. Interestingly, however, the high, thick clouds also make a relatively large contribution to the cloud forcing of albedo, particularly when compared to the albedo forcing by low clouds. High clouds contribute about 5.3% to the global albedo while covering about 18% of the surface area of Earth. Low clouds cover about 27% of the surface area and contribute only about 5.7% to the albedo, according to these estimates. In contrast to the estimates of Haurwitz (1948), Table 3 indicates that high clouds have typical albedos that are just as great as low clouds. The average cloud albedo is about the same for high, middle, and low clouds. This difference with Haurwitz's estimates is in large part a result of a difference in perspective. Haurwitz's cloud definitions were made by surface observers who could not see if higher clouds were above the low clouds and who defined cloudiness primarily by observing the base heights. Conversely, the ISCCP cloud definitions are based on the inferred cloud-top pressure and the visible optical depth, so that no information on cloud-base height is provided. Also, the estimates in Table 3 are an average constructed from nearly global data, whereas Haurwitz had data from only a single mid-latitude location. The global averages are heavily influenced by the tropical regions.

It is interesting that the high, thick clouds, which are mostly associated with tropical convective cloud complexes, have very high average albedos and produce a substantial reduction in net radiation. The longwave forcing by these clouds is large and positive, but this is overwhelmed by the powerful reduction of absorbed solar energy that these clouds produce. The decrease of net radiation by high, thick clouds per unit of area covered is actually larger than that of the average low cloud. Because they are much more abundant than high, thick clouds, however, low clouds have a larger net influence on the global energy budget. Also, high, thin clouds, which have a net positive effect on net radiation, occur in association with high, thick clouds and tend to mitigate their effect on the energy balance.

3. Effects of Clouds on Surface and Atmospheric Energy Budgets

Cloud radiative effects change the overall planetary energy budget and are important in determining the distribution of energy between the atmosphere and the surface (Stephens and Webster, 1984). Because the cloud-free atmosphere is relatively transparent to solar radiation and opaque to terrestrial radiation, the reflection of solar radiation by clouds affects the surface energy budget most directly, and the reduction in longwave cooling is felt primarily as a reduction in energy lost from the atmosphere. In this respect convective clouds are self-limiting to some extent, because their radiative effects reduce the radiative destabilization of the atmosphere that generates the need for convection. Ramanathan and Collins (1991) have argued that the high albedo of convective cloud complexes that occur in regions with warm sea-surface temperatures may play a critical role in preventing tropical sea temperatures from exceeding about 303 K. One of the important and not fully understood aspects of tropical cloudiness is that the presence or absence of deep convective clouds has very little influence on net radiation at the top of the atmosphere. Clouds also influence the internal heating profile within the atmosphere, primarily because cloud bases are heated by longwave energy fluxes while cloud tops are cooled. These radiative flux divergences help to sustain stratiform cloud decks near the surface and anvil clouds in the upper tropical troposphere, and they alter the large-scale radiative heating profile in the atmosphere (Houze, 1982).

B. Modeling Studies

Clouds and their radiative effects are predicted internally in many of today's climate models. Subtle changes in the amount, distribution, or radiative properties of clouds can have significant effects on the predicted climate. Studies have shown the great importance of cloud predictions both for the mean climate (e.g., Ramanathan et al., 1983; Slingo, 1990; Slingo and Slingo, 1988) and for the sensitivity of the climate to carbon dioxide increases (Hansen et al., 1984; Washington and Meehl, 1984; Wetherald and Manabe, 1988). Typically, these models predict that the cloud amount will decrease slightly and the average altitude of the cloud top will rise when the climate warms. If the albedos of the clouds are fixed, the enhanced greenhouse effect of these higher clouds dominates the cloud radiative feedback so that the cloudiness changes constitute a positive feedback.

Many climate models assume that cloud optical properties are fixed and consider only the feedbacks associated with cloud amount and altitude. Paltridge (1980) and Charlock (1981) considered how the sensitivity of a one-dimensional model would be affected if the cloud water content increases with temperature in proportion to the saturation vapor pressure. This produces a substantial negative cloud–climate feedback, except possibly for high, thin cirrus clouds. Somerville and Remer (1984) performed similar calculations but with the dependence of cloud water content determined from observations summarized by Feigelson (1978).

A substantial increase in sophistication of climate models would be to predict cloud optical properties as well as cloud distributions in space (Charlock and Ramanathan, 1985; Sundquist, 1993). Such models generally require the prediction of a physical cloud variable such as cloud liquid/ice water content, and the specification of an appropriate drop size distribution and cloud geometry. In models for which the cloud water amount increases with temperature, the increased cloud albedo forms a negative climate feedback (Roeckner et al., 1987). The total radiative effect of the liquid water increase on surface temperature is sensitive to its vertical distribution, however, and may in fact result in a positive feedback (Schlesinger, 1988). Longwave and shortwave radiative forcing by clouds have a substantial influence on the nonradiative surface fluxes (e.g., Betts and Ridgeway, 1988). Climate simulations that predict cloud water are very sensitive to the residence time of cloud water in the atmosphere after formation. The effect of temperature on the residence time through the phase changes of water has been proposed as an important feedback mechanism by Mitchell et al. (1989). They propose that as the climate warms, some of the water in the atmosphere is converted from ice to liquid, which has a longer residence time in the atmosphere. The resulting increase in mid-level cloud water constitutes a negative feedback.

Recent studies have pointed out the importance of cloud droplet radius and its relationship to aerosols that serve as cloud condensation nuclei. Increased production of aerosols may increase the available cloud condensation nuclei, so that the number of cloud droplets increases and their size decreases. The same amount of liquid water distributed over more droplets will produce a higher albedo (Charlson et al., 1987; Coakley et al., 1987; Radke et al., 1989). These smaller droplets may also persist longer in the atmosphere before precipitating. Changes in cloud condensation nuclei may be related to natural or anthropogenic production of sulfur-bearing gases. The aerosols resulting from anthropogenic sulfur gases may themselves cause significant perturbations in the radiation balance (Charlson et al., 1991).

On the basis of recent modeling work it is apparent that the radiative effects of clouds have a significant influence on the mean climate. Changes in cloud optical properties or distributions that may accompany a climate change could have a substantial effect on the magnitude and character of the climate change accompanying a doubling of atmospheric carbon dioxide, for example. While the cloud radiative forcing of the current climate can be somewhat successfully simulated with current climate models (Buriez et al., 1988; Cess and Potter, 1987; Harshvardhan et al., 1989; Kiehl and Ramanathan, 1990; Slingo, 1982), this constitutes no assurance that the cloud radiative feedbacks accompanying a climate change can be estimated with any confidence.

Recent intercomparisons of a large number of climate models with predicted clouds showed a wide variation in the effect of cloud feedback on climate sensitivity (Cess et al., 1990; Cess et al., 1989). The change in net radiation caused by an imposed sea-surface temperature change varied by a factor of three among these models, mostly as a result of differences in the treatment of the interaction of cloud radiative processes with climate. In addition, Kiehl and Williamson (1991) have shown that in a typical model, changes in spatial resolution can interact strongly with the cloudiness parameterization and thereby greatly change the mean climate of the model. It may be that the sensitivity of cloud amount to model resolution decreases when the model resolution becomes quite fine, but this must depend on the nature of the cloud parameterization.

IV.   Outstanding Problems

Recent advances in measuring the radiative energy budget and the cloud climatology of Earth have greatly enhanced our knowledge of how clouds affect the energy budget of the planet. The effects of clouds on the surface and top-of-atmosphere energy budgets are very significant, and depend sensitively on the cloud properties. Numerical models of climate also indicate that the mean climate of Earth is sensitive to cloud distributions and optical properties, and that changes in cloud properties can have dramatic effects on the response of the climate to perturbing influences such as increasing concentrations of atmospheric greenhouse gases.

A need exists to further improve estimates of cloud radiative forcing. Better methods of using satellite data to identify clouds and their effects over snow-covered surfaces need to be found. Much-improved estimates of water vapor concentrations are needed both to evaluate cloud radiative forcing and to validate parameterizations of moist processes in climate models. Water vapor profiles in the middle and upper troposphere are especially badly needed. To make substantial progress in modeling the role of clouds in climate sensitivity will require joining accurate estimates of cloud radiative effects with additional data on the physical properties of clouds such as particle size and the vertical and horizontal distribution of cloud water and ice.

Many possible cloud feedbacks have been suggested, which would make the climate more or less sensitive to changing greenhouse gases. Each of these feedbacks needs to be carefully evaluated through a combination of observational, theoretical, and modeling studies. If cloud feedback mechanisms are demonstrated to be of potential importance for climate change, they must be properly incorporated into climate models. Many of the likely sensitivity mechanisms depend on variables like cloud particle radius, water content, and cloud structure (Harshvardhan, 1982) that are not currently carried in global climate models. Moreover, if these variables were carried in global climate models it is not clear that we have the observational data base necessary to validate such advanced climate models.

Global data to validate such cloud-physics parameterizations in climate models can perhaps be obtained from advanced remote-sensing instruments on orbiting satellites, augmented by more detailed *in situ* data from aircraft.

Acknowledgments

The author's work on this subject has been supported by the NASA Earth Radiation Budget Experiment under contract number NAS1-18157. M. L. Michelsen prepared the color plates and G. C. Gudmundson provided editorial assistance.

References

Abbot, C. G., and Fowle, E. E. J. (1908). "Annals of the Astrophysical Observatory of the Smithsonian Institution, Vol. 2." Smithsonian Institution, Washington, D.C.

Aldrich, L. B. (1919). *Smiths. Misc. Coll.* **69**, 1–9.

Ångström, V. A. (1928). *Beitr. Phys. Atmos.* **14**, 8–20.

Ardanuy, P. E., Stowe, L. L., Gruber, A., Weiss, M., and Long, C. S. (1988). *J. Climate* **2**, 766–799.

Ardanuy, P. E., Stowe, L. L., Gruber, A., and Weiss, M. (1991). *J. Geophys. Res.* **96**, 18,537–18,549.

Arrhenius, S. (1896). *Phil. Mag.* Ser. 5, Vol. **41**, 237–276.

Barkstrom, B. R., and Smith, G. L. (1986). *Rev. Geophys.* **24**, 379–390.

Betts, A. K., and Ridgeway, W. (1988). *J. Atmos. Sci.* **45**, 522–536.

Bort, T. d. (1884). *Ann. de bureau central météorologique de France* t. iv, 2nd partie, 27.

Brunt, D. (1934). "Physical and Dynamical Meteorology." Cambridge University Press, Cambridge, England.

Buriez, J. C., Bonnel, B., Fouquart, Y., Geleyn, J. F., and Morcrette, J. J. (1988). *J. Geophys. Res.* **93**, 3,705–3,719.

Cess, R. D. (1976). *J. Atmos. Sci.* **35**, 1831–1848.

Cess, R. D., and Potter, G. L. (1987). *Tellus* **39A**, 460–473.

Cess, R. D., Briegleb, B. P., and Lian, M. S. (1982). *J. Atmos. Sci.* **39**, 53–59.

Cess, R. D., Potter, G. L., Blanchet, J. P., Boer, G. J., Ghan, S. J., Kiehl, J. T., Le Treut, H., Li, Z.-X., Liang, Z.-X., Mitchell, J. F. B., Morcrette, J.-J., Randall, D. A., Riches, M. R., Roeckner, E., Schlese, U., Slingo, A., Taylor, K. E., Washington, W. M., Wetherald, R. T., and Yagai, I. (1989). *Science* **245**, 513–516.

Cess, R. D., Potter, G. L., Blanchet, J. P., Boer, G. J., Déqué, M., Gates, W. L., Ghan, S. J., Kiehl, J. T., Le Treut, H., Li, Z.-X., Liang, X.-Z., McAveney, B. J., Meleshko, V. P., Mitchell, J. F. B., Morcrette, J.-J., Randall, D. A., Rikus, L., Roeckner, E., Royer, J. F., Schlese, U., Sheinin, D. A., Slingo, A., Sokolov, A. P., Taylor, K. E., Washington, W. M., Wetherald, R. T., and Yagai, I. (1990). *J. Geophys. Res.* **95**, 16,601–16,615.

Charlock, T. P. (1981). *J. Atmos. Sci.* **38**, 661–663.

Charlock, T. P., and Ramanathan, V. (1985). *J. Atmos. Sci.* **42**, 1408–1429.

Charlson, R. J., Lovelock, J. E., Andreae, M. O., and Warren, S. G. (1987). *Nature* **326**, 655–661.

Charlson, R. J., Langner, J., Rudhe, H., Leovy, C. B., and Warren, S. G. (1991) *Tellus* **43AB**, 152–163.

Coakley, J. A. J., Bernstein, R. L., and Durkee, P. A. (1987). *Science* **237**, 1020.

Danjon, A. (1936). *Ann. L'Obs. Strasbourg* **3**, 139–181.

Dhuria, H. L., and Kyle, H. L. (1990). *J. Climate* **3**, 1409–1434.

Ellis, J. S. (1978): "Cloudiness, the planetary radiation budget and climate." Colorado State University, Colorado.

Feigelson, E. M. (1978). *Contrib. Atmos. Phys.* **51**, 203–229.

Fouquart, Y., Buriez, J. C., Herman, M., and Kandel, R. S. (1990). *Rev. Geophys.* **28**, 145–166.

Fritz, S. (1949). *J. Meteor.* **6**, 277–282.

Hansen, J., Lacis, A., Rind, D., Russell, G., Stone, P., Fung, I., Lerner, J., and Ruedy, R. (1984). (J. E. Hansen and T. Takahashi, eds.), In "Climate Processes and Climate Sensitivity," 130–163. AGU, Washington DC.

Harrison, E. F., Minnis, P., Barkstrom, B. R., Ramanathan, V., Cess, R. D., and Gibson, G. G. (1989). *J. Geophys. Res.* **94**, 18,687–18,703.

Harshvardhan, J. (1982). *J. Atmos. Sci.* **39**, 1853–1861.

Harshvardhan, J., Randall, D. A., Corsetti, T. G., and Dazlich, D. A. (1989). *J. Atmos. Sci.* **46**, 1922–1942.

Hartmann, D. L., and Short, D. A. (1980). *J. Atmos. Sci.,* **37**, 1233–1250.

Hartmann, D. L., and Doelling, D. E. (1991). *J. Geophys. Res.* **96**, 869–891.

Hartmann, D. L., Ockert-Bell, M. E., and Michelsen, M. L. (1992). *J. Climate* **5**, 1281–1304.

Hartmann, D. L., Ramanathan, V., Berroir, A., and Hunt, G. (1986). *Rev. Geophys.* **24**, 439–468.

Haurwitz, B. (1948). *J. Meteor.* **5**, 110–113.

Houghton, H. G. (1954). *J. Appl. Meteor.* **11**, 1–9.

House, F. B., Gruber, A., Hunt, G. E. and Mecherikunnel, A. T. (1986). *Rev. Geophys.* **24**, 357–377.

Houze, R. A. (1982). *J. Meteor. Soc. Japan* **60**, 396–410.

Hunt, G. E. (1982). *Tellus* **34**, 29–38.

Hunt, G. E., Kandel, R., and Mecherikunnel, A. T. (1986). *Rev. Geophys.* **24**, 351–356.

Hwang, P. H., Stowe, L. L., Yeh, Y. M., Kyle, H. L., and N. 7. C. D. P. Team (1988). *Bull. Amer. Meteor. Soc.* **69**, 743–752.

Jacobowitz, H., Soule, H. V., Kyle, H. L., House, F. B., and Nimbus-7-Team (1984). *J. Geophys. Res.* **89** (D4), 5021–5038.

Kiehl, J. T., and Ramanathan, V. (1990). *J. Geophys. Res.* **95**, 11,679–11,698.

Kiehl, J. T., and Williamson, D. L. (1991). *J. Geophys. Res.* **96**, 10,955–10,980.

King, M. D. (1993). (P. V. Hobbs, ed.), *In* "Cloud-Aerosol-Climate Interactions" Academic Press, San Diego.

Kyle, H. L., Ardanuy, P. E., and Hurley, E. J. (1985). *Bull. Amer. Meteor. Soc.* **66**, 1378–1388.

London, J. (1956). "A study of the atmospheric heat balance." Final Report, Contract AF19(122). New York University, Research Division, College of Engineering.

Luckiesh, M. (1919). *Astrophys. J.* **69**, 108–130.

Manabe, S., and Strickler, R. F. (1964). *J. Atmos. Sci.* **21** (4), 361–385.

Manabe, S., and Wetherald, R. T. (1967). *J. Atmos. Sci.* **24**, 241–259.

Meleshko, V. P., and Wetherald, R. T. (1981). *J. Geophys. Res.* **86**, 11,995–12,014.

Mitchell, J. F. B., Senior, C. A., and Ingram, W. J. (1989). *Nature* **341**, 132–134.

Neiburger, M. (1949). *J. Meteor.* **6**, 98–104.

Ohring, G., and Clapp, P. F. (1980). *J. Atmos. Sci.* **37**, 447–454.

Ohring, G., and Gruber, A. (1983). *Adv. Geophys.* **25**, 237–304.

Ohring, G., Clapp, P. F., Heddinghaus, T. R. and Krueger, A. F. (1981). *J. Atmos. Sci.* **38**, 2539–2541.

Paltridge, G. W. (1980). *Quart. J. Roy. Meteor. Soc.* **106**, 367–380.

Radke, L. F., Coakley, J. A. J., and King, M. D. (1989). *Science* **246**, 1146–1149.

Ramanathan, V. (1987). *J. Geophys. Res.* **92**, 4075–4095.

Ramanathan, V., and Collins, W. (1991). *Nature* **351**, 27–32.

Ramanathan, V., Pitcher, E. J., Malone, R. C., and Blackmon, M. L. (1983). *J. Geophys. Res.* **40**, 605–630.

Ramanathan, V., Cess, R. D., Harrison, E. F., Minnis, P., Barkstrom, B. R., Ahmad, E., and Hartmann, D. L. (1989). *Science* **243**, 57–63.

Raschke, E., Vonder Haar, T. H., Bandeen, W. R., and Pasternak, M. (1973). *J. Atmos. Sci.* **30**, 341–364.

Roeckner, E., Schlese, U., Biercamp, J., and Loewe, P. (1987). *Nature* **329**, 138–140.

Rossow, W. B., and Schiffer, R. A. (1991). *Bull. Amer. Meteor. Soc.* **72**, 2–20.

Schiffer, R. A., and Rossow, W. B. (1983). *Bull. Amer. Meteor. Soc.* **64**, 779–784.

Schlesinger, M. E. (1988). *Nature* **335**, 303–304.

Schmetz, J., Mhita, M., and Van de Berg, L. (1990). *J. Climate* **3**, 784–791.

Schneider, S. H. (1972). *J. Atmos. Sci.* **29**, 1413–1422.

Slingo, A. (1990). *Nature* **343**, 49–51.

Slingo, A., and Slingo, J. M. (1988). *Q.J.R.M.S.* **114**, 1027–1062.

Slingo, J. (1982). *Q.J.R.M.S.* **108**, 379–405.

Smagorinsky, J. (1963). *Mon. Wea. Rev.* **91**, 99–164.

Smagorinsky, J., Manabe, S., and Holloway, J. L. (1965). *Mon. Wea. Rev.* **93**, 727–768.

Smith, G. L., Green, R. N., Raschke, E., Avis, L. M., Suttles, J. T., Wielicki, B. A. and Davies, R. (1986). *Rev. Geophys.* **24**, 407–421.

Somerville, R., and Remer, L. A. (1984). *J. Geophys. Res.* **89**, 9668–9672.

Stephens, G. L., and Webster, P. J. (1984). *J. Atmos. Sci.* **41**, 681–686.

Stephens, G. L., Campbell, G. G., and Vonder Haar, T. H. (1981). *J. Geophys. Res.* **86**, 9739–9760.

Stowe, L. L., Yeh, H. Y. M., Eck, T. F., Wellemeyer, C. G., Kyle, H. L., and N.-7. C. D. P. Team (1989). *J. Climate* **2**, 671–709.

Sundquist, H. (1993). (P. V. Hobbs, ed.), *In* "Aerosol-Cloud-Climate Interactions," Academic Press, San Diego.
Vonder Haar, T. H., and Suomi, V. E. (1971). *J. Atmos. Sci.* **28**, 305–314.
Washington, W. M., and Meehl, G. A. (1984). *J. Geophys. Res.* **89**, 9475–9503.
Wetherald, R. T., and Manabe, S. (1980). *J. Atmos. Sci.* **37**, 1485–1510.
Wetherald, R. T., and Manabe, S. (1988). *J. Atmos. Sci.* **45**, 1397–1415.
Wielicki, B. A., and Green, R. N. (1989). *J. Appl. Meteor.* **28**, 1133–1145.

Chapter 7 | Parameterization of Clouds in Large-Scale Numerical Models

Hilding Sundqvist
Department of Meteorology
Stockholm University
Stockholm, Sweden

General aspects of deducing a consistent cloud parameterization scheme, designed to predict fractional cloud cover and precipitation amounts and distribution in large-scale models, are discussed. For specific background conditions, such as the supply of cloud condensation nuclei and large-scale and mesoscale circulations, the development of hydrometeors is governed by cloud microphysical processes. Cloud cover is determined by humidity and air motions, resulting from both the larger-scale background circulation and the release of latent heat on smaller scales. Owing to nonlinearities, the parameterizations of these two widely separated scale regimes are not completely decoupled. However, the dominating mechanisms of each regime may be considered separately. Thus, the basic question in parameterization of the microphysics is how to partition the condensate, which is produced by condensation, into cloud and precipitating particles (liquid and solid). The basic problem regarding parameterization of the meso- (subgrid-) scale motion—that is, of fractional cloud cover—is to partition the water vapor, which is available in a grid box, between condensation and moistening in the box.

For microphysical processes, it is concluded that different schemes should be employed for liquid clouds and ice crystal (cirrus) clouds as well as for mixed liquid–ice clouds because of the different growth characteristics of the particles. For the parameterization of mesoscale cloud cover, it is necessary to provide different descriptions for convective and stratiform clouds due to different types of energy conversions. For stratiform clouds, special attention must be paid, for example, to stratocumulus and cirrus clouds.

Questions concerning data needs are also addressed. The discussion on data considers the requirements for development, verification, and validation of parameterization schemes, and for establishment of initial cloud water and humidity fields.

I. Introduction

Clouds form as a result of atmospheric condensation, which interacts with the atmospheric mass and wind fields through liberated heat. Clouds have a strong influence on the transmission of shortwave and longwave radiation in the atmosphere. Clouds also have important effects on the atmosphere on both short and long time scales. An example of a short-range effect is the influence of cloudiness on the diurnal variation of the structure of the atmospheric boundary layer. However, it is the long-term cloud-radiation interactions—monthly, seasonal, and even longer periods—that have received most attention. This interest is prompted by the need to acquire insight into the climate state and to assess possible anthropogenic impacts on climatic change (see, e.g., Ramanathan et al., 1989; Fouquart et al., 1990; IPCC, 1990; Ramanathan and Collins, 1991).

During the last decade, a great number of simulation studies of the general circulation and climate variations have been performed with the aid of GCMs (general circulation models). Practically without exception, the results show a pronounced sensitivity to the way cloudiness is treated in the models (e.g., Ramanathan et al., 1983; Cess et al., 1989; Röckner and Schlese, 1987; Wetherald and Manabe, 1988; Slingo and Slingo, 1991). Because of insufficient knowledge about cloud processes, they are handled differently in various models. Consequently, conclusions based on GCMs suffer from an unsatisfactorily, if not unacceptably, large uncertainty. The need for improved cloud parameterization has been explicitly emphasized in several recent studies (Wetherald and Manabe, 1988; Randall et al., 1989; IPCC, 1990).

Another important class of applications, in which a realistic description of cloud processes is crucial, is modeling of atmospheric chemistry. Realistically simulated chemical reactions and transformations in forming and reevaporating clouds and precipitation are vital prerequisites for meaningful studies of the distribution and deposition mechanisms of various chemical compounds in the atmosphere. An example of this type of application is found in Berge (1990). A study of the role of clouds in tropospheric photochemistry may be found in Lelieveld and Crutzen (1991).

The present-day numerical models designed for short- or medium-range numerical weather prediction (NWPM) are generally quite elaborate in their treatment of vertical eddy fluxes, radiative transfer, and energy fluxes at the earth's surface, particularly over land areas. However, cloud processes are not included in the models with the same degree of refinement. Very few studies have been performed to scrutinize whether or not the cloudiness treated by NWPM (or GCMs) is realistic. Therefore, we have very little insight into the validity of the cloud parameterizations. This situation can be explained partially by the fact that it has been difficult to obtain relevant data for quantitative verification of the model clouds. Yet, with regard to the strong impact that clouds have on the model results, it appears absolutely necessary that cloudiness be simulated with greater certainty and accuracy if the model results are to be used with any confidence, particularly for studies of climate sensitivity (Ramanathan et al., 1989; Slingo, 1990).

Except for a few models that treat cloud water content (Röckner and Schlese, 1987; Mitchell et al., 1989; Slingo and Slingo, 1991), most NWPM and GCM of today merely consider cloud cover. The treatment of cloud cover is generally characterized by *ad hoc* approaches. Consistency between volumes of model condensation and of model cloud field is not always guaranteed; for example, cloud water is considered only when it is produced by stratiform condensation, and advection of the water is omitted. Smith (1990) presents an elaborate discussion, not only on inclusion of fractional cover, but also on incorporation of cloud water as a prognostic variable.

The purpose of this review is to discuss the basic elements of a consistent parameterization of the closely linked condensation–cloud–precipitation–evaporation processes. The topic is, in its full extent, very complex and cannot be covered comprehensively here. Therefore, this discussion will focus on the question of parameterizing the *entity cloud*—that is, the physical entity consisting of part of the condensate and occupying a certain volume of air that is often only a fraction of a grid volume. Elaborate considerations will be left out, but references to special studies will be made. It is not feasible to present a complete list of relevant references, but many of the referred papers contain references that are supplementary in this context.

II. Parameterization of Cloud Processes

A. General

Fractional cloud cover is commonly the only cloud parameter that is considered in models containing radiation calculations. Cloud water content is taken into account in only a few models. We know that the water path of clouds is important for the radiation budget (e.g., Stephens and Webster, 1981). Therefore, since a cloud is made up of water, liquid or ice, it appears natural that a minimum requirement for a more elaborate treatment of clouds in numerical models is to include cloud water content as a dependent variable in the model set of equations. When we have learned to simulate cloud water reliably, it may become meaningful to look into the effects of cloud condensation nuclei on droplet size distributions and the cloud albedo (e.g., Charlson et al., 1987). Since cloud water is a measurable quantity, it is possible to quantitatively verify simulated amounts and distributions.

Some approaches to accounting for cloud water content use a diagnostic parameterization based on statistical models (Sasamori, 1975; Hense and Heise, 1984). Sommeria and Deardorff (1977) suggested a statistical/empirical approach that accounts for both cloud cover and cloud water content. Their concept has been adopted by Smith (1990) in a fairly comprehensive scheme for layered clouds. In the present discussion, we will consider a parameterization for cloud water that is based on the physical mechanisms involved in the processes of condensation and evolution of the condensed matter.

The inclusion of cloud water as a prognostic variable implies that the regime of microphysical processes has to be taken into account in the parameterization. This regime includes the description of the condensation itself, the growth of cloud particles to precipitating size, and the evaporation of precipitation during its fall. The parameterization of condensation also consists of another regime. This is composed of the mesoscale circulation that is induced by the released heat of condensation. With regard to commonly employed model resolutions, this circulation is subgrid-scale, hence leading to the appearance of fractional cover in grid boxes. We will start from the most generally valid formulation of the parameterization problem and then discuss more detailed treatments of involved scales and physical mechanisms.

Treatises and books on cloud physics have provided insight and knowledge of the microphysical aspects of the condensation process (e.g., Mason, 1971; Pruppacher and Klett, 1978). However, only a few studies have been performed where those processes are considered in conjunction with the mesoscale processes that are responsible for cloud formation (e.g., Rutledge and Hobbs, 1983; Cotton and Anthes, 1989). Thus, we begin with the general problem of deciding how to specify where condensation will take place; what will be the rate of latent heat release, and consequently the production rate of condensate; how large a fraction of the grid box will be occupied by these clouds; and whether or not any evaporation of hydrometeors will take place.

We have already noted one plausible reason why the inclusion of elaborate cloud treatments in numerical models has been so slow. Another equally important reason is probably the additional computation time that follows such an inclusion. It has been considered impracticable to introduce lengthy calculations of microphysical processes. Therefore, in the present discussion of approaches, we will compromise somewhat between generality and practicability. We will consider examples of simplifications, which are believed to yield tractable schemes for large-scale models. However, although we reduce the generality or completeness to some extent, the discussion still gives a fairly complete demonstration of the principal aspects of the problem of parameterizing condensation in clouds.

We now proceed to discussions of the microscale and the mesoscale parameterization regimes separately. However, because of nonlinear relations, we will find that it will also be necessary to consider them together.

B.   Condensation

As a result of theoretical studies, laboratory experiments, and field studies by cloud physicists, we have a fair understanding of the microphysical processes that take place in condensation and in the subsequent growth of condensate in clouds. It is this insight, mainly concerning individual particles, that we must use when we deduce formulations to be valid for ensembles on the scale of a cloud. The principles behind this approach are demonstrated in Kessler's (1969) paper.

Since cloud water content is the most important hydrometeor to be included in a more elaborate treatment of clouds in large-scale models, this will be the only *additional prognostic variable* in the prediction system to be discussed in this section. Although we will not include prognostic equations for other types of hydrometeors, the generality of the discussion is not significantly reduced. Rather, as a consequence, we are confronted with both the explicit and indirect parameterization problems that arise because some of the needed quantities are not available. When other related quantities are required, such as the density of precipitating water and distinctions between liquid and solid phases, they have to be deduced diagnostically. The latter aspect is the reason for using just "water content" and not "liquid water content" in the present text. In this section, we will deal with in-cloud conditions, implying, for example, that cloud water content is the local mixing ratio and not a (mean) value representative of a gridpoint. To make our discussions concrete, we consider the tendency equations for temperature T, specific humidity (interchangeably called moisture) q, and cloud water mixing ratio m:

$$\frac{\partial \hat{T}}{\partial t} = A_{\hat{T}} + \frac{L}{c_p} \hat{Q} \tag{1}$$

$$\frac{\partial \hat{q}}{\partial t} = A_{\hat{q}} - \hat{Q} \tag{2}$$

$$\frac{\partial \hat{m}}{\partial t} = A_{\hat{m}} + \hat{Q} - \hat{G}_P \tag{3}$$

where the in-cloud values are indicated by the hat symbol, $\hat{}$. The A terms contain all tendency contributions except those connected with the condensation process. The latent heat of condensation is denoted by L and the specific heat at constant pressure by c_p. The rate of latent heat release, or rate of production of condensate, is $\hat{Q}$ (dimension s^{-1}), and the rate of generation of precipitation is $\hat{G}_P$. As indicated, we deal with cloud water as if it is liquid; we will discuss the distinction between liquid and ice later.

Empirically we know that condensation takes place at very low supersaturations as a result of cloud condensation nuclei in the air. Hence, at the present state of the art, it seems justifiable to assume that condensation occurs at 100% relative humidity; of course, vapor must be available, either through convergence into the volume in question or through cooling by expansion or radiation. The rate of condensation is given by the rate at which vapor is made available; the formulation is derived from Eqs. (1) and (2).

Note that $\hat{q}$ is the saturation value, $q_s(\hat{T})$. Take the partial time derivative of the latter, which consists of a pressure tendency and a temperature tendency through the Clausius–Clapeyron relation. Let this expression replace the left-hand side of Eq. (2). Then eliminate the temperature tendency with the aid of Eq. (1) to obtain, after rearrangement, the rate of production of condensate:

$$\hat{Q} = \left[A_{\hat{q}} - \frac{c_p}{L} S_q A_{\hat{T}} + \frac{\hat{q}}{p} \frac{\partial p}{\partial t} \right] (1 + S_q)^{-1}$$

$$S_q = \frac{L}{c_p} \frac{q_s}{E_s} \frac{dE_s}{dT} = \frac{\epsilon L^2}{Rc_p} \frac{q_s}{T^2}$$

(4)

where p is pressure, E_s is the saturation vapor pressure, R is the gas constant of dry air, and $\epsilon = 0.622$. (Further details of the derivation may be found in Sundqvist, 1988.) For condensation to occur, the numerator of Eq. (4) *must be positive*; furthermore, we see that vapor may be made available not only through flux convergence, but also through cooling and pressure changes.

It is less clear what the saturation condition for phase change at low temperatures should be. Freezing nuclei may be few in the upper troposphere. Therefore, cirrus crystal nucleation is accomplished by homogeneous freezing, which may take place at temperatures of about $-37°C$ for solution droplets (Heymsfield and Sabin, 1989). Then saturation with respect to water should be applied. However, after a cirrus cloud has formed, ice crystals may act as freezing nuclei, and saturation with respect to ice will then be appropriate. (The problem is no easier at intermediate temperatures where supercooled water and ice coexist; see for example Rangno and Hobbs, 1991). A specific application is found in Heymsfield and Donner (1990), who suggest a diagnostic scheme for the derivation of the amount of ice cloud water in GCM. In a more detailed cirrus cloud model, Starr and Cox (1985) employed a relative humidity condition that varied between 120 and 100%, depending on whether or not a cloud existed and the position in the cloud (inside or at the top).

C. Release of Precipitation

One of the main goals for condensation–cloud parameterization is to describe the partitioning of condensate into cloud water and precipitating water. The various microphysical processes that are active in this context are basically the same, whether they take place in stratiform or convective clouds. As a result of differences in vertical velocity between these two stability regimes, the participating microphysical processes may be differently emphasized in the two cases. This circumstance should be reflected in the magnitude of the parameters of the deduced parameterization scheme. On the other hand, there are distinct differences in growth characteristics among all ice clouds, all water clouds, and mixed ice–water clouds. Hence, it is necessary to derive different parameterization schemes for these cloud types. This aspect will be considered at the end of this subsection.

In his approach, Kessler (1969)—who considered only liquid water clouds—included prognostic equations for both cloud water and precipitating water, implying that he could derive explicit formulations not only for "autoconversion" but also for "collection of cloud water by rain." In the first process, cloud droplets in their random motion occasionally collide and coalesce. In the second process, raindrops, falling through a cloud, grow by collecting cloud droplets. However, in the present discussion, only cloud water is available. Thus, only autoconversion can be described explicitly. The implications of this are discussed in the next few paragraphs.

To describe the release of precipitation, Sundqvist (1978) suggested a continuous nonlinear function of cloud water of the form:

$$\hat{G}_P = C_0 \cdot \hat{m} \cdot f\left(\frac{\hat{m}}{m_r}\right) \tag{5}$$

where C_0 and m_r are the basic parameters. For $(\hat{m}/m_r) = x < 1$, the function $f(x) \ll 1$, but f tends to unity as x increases; hence not until $\hat{m}$ is of about the same magnitude as m_r does the cloud generate precipitation with some efficiency. The low generation rate when $x < 1$, accounts for the slow diffusive growth of small droplets. Kessler (1969) simulated this effect by introducing a minimum cloud-water amount that has to be exceeded before the autoconversion becomes active.

To include the effect of coalescence by rain, it is necessary to know the density of the precipitating mass. The derivation starts with the growth rate of a raindrop, which is proportional to the cloud water content and to the terminal velocity of the drop. Integrating over the drop size distribution (e.g., Marshall and Palmer, 1948), a growth rate that is a function of precipitating mass is obtained (Kessler, 1969). In our case, the rain water content first has to be deduced. A diagnostic value of this is obtained by dividing the rate of precipitation by a typical terminal velocity; this is proportional to the rain water content raised to a small power, about 1/8, implying that the velocity varies relatively little even for large variations in water content (Kessler, 1969). Consequently, Sundqvist et al. (1989) accounted for this coalescence effect by taking:

$$C_0 = C_{00}F_{co}; \qquad F_{co} = 1 + C_1\tilde{P}^{1/2} \tag{6}$$

and

$$\tilde{P}(p) = \frac{1}{g} \int_{p\text{-top}}^{p} \hat{G}_P \, dp$$

where g is the acceleration of gravity and C_1 is an additional parameter. Inserting the expression for C_0 in Eq. (5) we get two terms expressing the release of precipitation—one due to autoconversion and one resulting from collection by rain—similar to what one obtains (more naturally) when both cloud water and precipitation water are considered explicitly (Kessler, 1969). In Kessler's derivation, the rain water content is raised to the power 7/8 in the term representing collection by rain. In Eq. (6), the power 1/2 was adopted to enhance the effect at small precipitation rates and hence to compensate for the reduction by the factor f, which is small for small m and hence for low precipitation rates. To further curb this diminishing effect of f, m_r is divided by F_{co}. Golding (1986) adds the corresponding coalescence term without the factor f, and omits exponentiation of the precipitation rate. (In a strict transformation of Kessler's expression where the precipitation density is replaced by the rate of precipitation, the latter should be raised to 7/9, because the mass-averaged terminal velocity is proportional to the precipitation density raised to the power 1/8.) Flatøy (1992) found that the Kessler approach and the scheme of Sundqvist et al. (1989) give very similar results for appropriate values of C_{00} and C_1 in Eq. (6).

In their model experiment, Sundqvist et al. (1989) first imposed a restriction on the role of coalescence ($F_{co} < 5$), whereby the resulting vertically integrated cloud water content even exceeded 40 mm in the most extreme areas; when this restriction was removed, the model produced values of vertically integrated cloud water content generally smaller than 2 mm, and no value above 5 mm was found in the simulation. Hence, these results indicate that it is necessary to include the coalescence effect by rain to obtain a realistic partitioning of the condensate between cloud water and precipitating water.

To obtain a realistic picture of the release of precipitation—and consequently a realistic distribution of cloud water—it seems important to account for the so-called Bergeron–Findeisen mechanism (Bergeron, 1935; Findeisen, 1938), which arises in a cloud where ice crystals and water droplets coexist. In this situation, the difference in saturation vapor pressure over water and ice causes a fast growth of the ice particles at the expense of the droplets. This mechanism was for a long time considered the major (or even the only) cause for release of precipitation. A mixed ice–water state appears when supercooled droplets freeze and the supply of droplets is replenished through condensation at a rate as fast as, or faster than, the freezing process. This heterogeneous freezing is brought about by the freezing nuclei available in the volume in question, or/and by precipitating ice crystals that have formed in layers above. Consequently, the Bergeron–Findeisen effect exists in the temperature interval 232–273 K. The altitude (or temperature) at which the maximum effect appears is determined by the above-mentioned saturation vapor difference, the efficiency of available freezing nuclei, and the vertical distribution of ice/snow precipitation. Thus, to take the Bergeron–Findeisen mechanism into account, it becomes necessary to consider the probability for existence or creation of ice crystals. Smith (1990) infers ice crystal existence from temperature to simulate the Bergeron–Findeisen effect. We note that Rangno and Hobbs (1991) found that there is no unique relation to temperature alone for the existence of ice crystals in mixed clouds. From extensive empirical data on cloud water, Matveev (1984) argues that temperature is the prime indicator of ice probability in clouds.

Below about -37 or $-40°C$, most clouds consist only of ice crystals. Then neither accretion nor Bergeron–Findeisen processes are active, implying that the particles grow essentially by deposition of vapor or aggregation. In this case the partitioning of the condensate into cloud water and precipitation requires a different description than for liquid clouds. In their cirrus cloud model, with ice water content as a prognostic variable, Starr and Cox (1985) incorporated an elaborate parameterization of the microphysical processes. Such approaches may provide valuable insights that can be utilized for development of less detailed schemes, which are suited for large-scale models. For application in GCM, Heymsfield and Donner (1990) derived a simpler scheme in which ice water was treated essentially as a diagnostic quantity.

D. Evaporation and Melting

Evaporation will take place when hydrometers are brought into subsaturated air volumes. *Cloud water* does not survive long when the air is subsaturated, due to the small size of cloud particles (Mason, 1971; Pruppacher and Klett, 1978). Therefore, in view of the typical length of a time-step in numerical models, it seems reasonable to assume that cloud water evaporates instantly when it enters a subsaturated volume.

Evaporation of *precipitating water* proceeds at a slower rate, which means that a more detailed derivation is required for its parameterization. We adopt the Marshall–Palmer (1948) drop size distribution. The rate equation for evaporation of a single drop (Mason, 1971; Pruppacher and Klett, 1978) is then integrated over the size spectrum to produce the rate of evaporation of the precipitating water mass, which becomes proportional to the square root of the density of the precipitating water. Kessler's (1969) derivation yields an exponent of 13/20, because he incorporates ventilation effects. Since we do not have the precipitating water explicitly available in the present approach, it is necessary to make the same diagnostic interpretation as when we formulated the coalescence effect in Eq. (6). Hence, we get for the evaporation rate of precipitating water:

$$\hat{E}_p = k_E(U_s - U)\tilde{\hat{P}}^{1/2} \tag{7}$$

where U is relative humidity and $U_s \equiv 1$.

For precipitating ice particles, it is important that saturation with respect to ice be applied to obtain the commonly observed long fall distances of ice particles (Starr and Cox, 1985), which in turn may be important for enhancement of the release of precipitation in liquid clouds below. Heymsfield and Donner (1990) deduced a scheme for the evaporation of ice crystals based on a detailed consideration of crystal size, fallspeed, and saturation with respect to ice. Their scheme demonstrates the basic questions that have to be dealt with in ice phase parameterization. The scheme involves an integration over a high-resolution representation in each model layer. It is likely that a further parameterization will be needed to account for this high-resolution integration so that the computational work is reduced.

When the ice phase is treated, it is also necessary that *melting* be taken into account at temperatures higher than 273 K. We perform a derivation similar to the one above for the rate of evaporation. Instead of subsaturation we now have a temperature excess, so we obtain (Mason, 1971):

$$\hat{S}_{melt} = k_{melt}(T - T_0)\tilde{P}^{1/2} \quad \text{for} \quad T > T_0, \quad T_0 = 273 \text{ K} \tag{8}$$

The melting coefficient k_{melt} will have a strong influence on the ratio of snow/rain in the precipitation that reaches the ground.

E. Parameters

In the above discussion, no reference was made to any specific model resolution. This means that the parameters of the microphysical schemes are merely connected with a particular type of condensation (water or ice) and cloud (convective or stratiform). Naturally, we have to utilize information from observations combined with results from model experimentation to determine the values that should be assigned to the parameters; these values should then be valid for arbitrary grid resolutions.

The values that have most commonly been used for the parameters appearing in Eqs. (5)–(8) are given in Table 1.

F. Parameterization of Subgrid-Scale Features

The amount of vapor available for possible condensation in a grid box is primarily given by quantities on the resolved scale (the gridpoint values), such as convergence of vapor and cooling. The appearance of subgrid-scale condensation—that is, the occurrence of fractional cloud cover—implies that the gridpoint relative humidity has not reached 100% $(U < 1)$. Therefore, the basic problems of subgrid-scale parameterization are

Table 1
Parameter Values for Microphysical Processes

Parameter	SI units	Convective case	Stratiform case
C_{00}	s^{-1}	10^{-3}	10^{-4}
m_r	$kg\,kg^{-1}$	$5 \cdot 10^{-4}$	$3 \cdot 10^{-4}$
C_1	$s^{-1}\,kg^{-1/2}\,m^{-1}\,s^{-1/2}$	300	300
k_E	$s^{-1}\,kg^{-1/2}\,m^{-1}\,s^{-1/2}$	10^{-5}	10^{-5}
k_{melt}	$s^{-1}\,K^{-1}\,kg^{-1/2}\,m^{-1}\,s^{-1/2}$	$5 \cdot 10^{-5}$	$5 \cdot 10^{-5}$

- to describe when condensation may occur (e.g., at relative humidity greater than a threshold value);
- To partition the converging vapor into one part that condenses and one part that changes the relative humidity.

This partitioning is the net result of a generally complex subgrid-scale circulation. The circulation is the interactive product of the basic flow on the resolved scale and the subgrid- (meso-) scale circulation, which follows from the condensational heating. Ideally, we need to understand how the cloud-connected circulation is related to the circulation on the resolved scale. Then we would be able to describe how large a fraction of the grid box is occupied by cloud and to formulate how the relative humidity behaves on the subgrid-scale. Current insight into these questions is insufficient for deducing parameterization relations of acceptable validity. A number of observational and experimental studies on mesoscale motion have been performed during the last decade (e.g., Browning, 1985, 1990; Locatelli and Hobbs, 1987). These studies have contributed to an enhanced qualitative insight into the dynamic interplay in and around atmospheric circulation systems on the mesoscale and the synoptic scale. However, we have not yet reached the stage where we have quantitative formulations for these processes that can be used in large-scale models. These basic questions concern both convective and stratiform condensation.

Strictly, this discussion should consider fractional occupation of a volume. But the conditions governing the vertical and horizontal extensions of a cloud follow from different physical processes. Therefore, in what follows, we will consider vertical and horizontal cloud cover separately and let the notation *fractional* imply the horizontal coverage and consequently indirectly assume that a cloud fills the whole space between two model levels. Later we shall also pay some attention to the question of vertical subgrid-scale situations (inversion clouds).

The gridpoint value of the model variables is a weighted mean of the subgrid scale features in a grid box. Figure 1 is a sketch of the condensation situation we are considering. The fractional cloud cover in the box is b, where the average value of in-cloud variables is denoted by $\hat{X}$ and the relative humidity is assumed

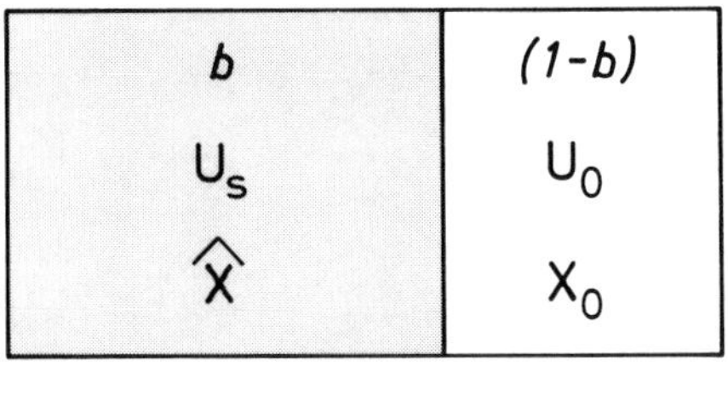

b	$(1-b)$
U_s	U_0
$\hat{X}$	X_0

Figure 1 The main structure of subgrid-scale condensation. Fraction b of the box, where $U = U_s \equiv 1$, is occupied by cloud, and fraction $(1 - b)$ is cloudfree with a mean relative humidity $U = U_0$.

to be 100% ($U = U_s \equiv 1$). In the cloud-free part, the corresponding quantities are X_0 and $U = U_0$. The gridpoint value of X is then

$$X = b \cdot \hat{X} + (1 - b) \cdot X_0 \tag{9}$$

Specifically we have

$$U = b \cdot U_s + (1 - b) \cdot U_0 \tag{10a}$$

$$m = b \cdot \hat{m} \tag{10b}$$

$$G_P = b \cdot \hat{G}_P = C_{00} \cdot m \cdot f\left(\frac{m}{b \cdot m_r}\right) \tag{10c}$$

$$+ C_{00} \cdot C_1 \cdot m \cdot \left(\frac{\tilde{P}}{b}\right)^{1/2} \cdot f\left(\frac{m}{b \cdot m_r}\right)$$

Equation (10a) may be rearranged to give an explicit expression for the cloud cover b in terms of relative humidity. This relation, which often is referred to as "a simplistic relation between relative humidity and cloud cover" follows from a straightforward averaging rather than from any simplifying assumptions about the physics involved. However, the seemingly simple relation contains the problem of subgrid-scale parameterization. Namely, we must know how to describe the subgrid-scale feature—the relative humidity U_0—in the cloud-free portion of the gridbox. At the moment when condensation begins, U_0 has the clear meaning of a threshold value. Then, as condensation continues, a formulation is needed for the rate of change of U_0. Hence, this is another expression for the partitioning problem mentioned at the beginning of this section. We will return to these questions later in discussing condensation in stratiform and convective clouds.

Equations (10b,c) exhibit another important point that must be taken into account in the subgrid-scale parameterization. Because of the nonlinearity of Eqs. (5) and (6), cloud cover appears in the expression of microphysical parameterization, Eq. (10c); this is important, because it is in this way that we use the general validity of the microphysical parameters discussed in Section II.E. This feature does not seem to have been discussed or considered in some of the GCMs that have used Kessler's (1969) formulations of autoconversion and evaporation of rain, which are derived for in-cloud conditions.

We now turn to the closure problem presented by the thermodynamic and moisture model equations. Applying the averaging procedure demonstrated by Eq. (9), those equations become

$$\frac{\partial T}{\partial t} = A_T + \frac{L}{c_p} Q \tag{11}$$

$$\frac{\partial q}{\partial t} = A_q - Q \tag{12}$$

In the interest of clarity, we let Q denote the net heating from the condensation–

evaporation taking place in the grid box (a more detailed treatment with explicit splitting into condensation and evaporation can be found in Sundqvist et al., 1989). (The resulting equations contain some approximations because the local tendency and horizontal gradients of cloud cover have been neglected.) Equations (11) and (12) have the same form as Eqs. (1) and (2), but there is an important difference in that q in Eq. (12) is not the saturation value. Therefore, to derive an equation corresponding to Eq. (4), we must take into account that $q = U \cdot q_s(T)$. Then, the equation corresponding to Eq. (4) is

$$Q = \left[A_q - \frac{c_p}{L} S_q A_T + U q_s \frac{1}{p} \frac{\partial p}{\partial t} - q_s \frac{\partial U}{\partial t} \right] (1 + U S_q)^{-1}$$

$$= \left[M_q - q_s \frac{\partial U}{\partial t} \right] (1 + U S_q)^{-1}$$

(13)

where S_q is defined in Eq. (4). Equation (13) shows that to close our system it is necessary to deduce a formulation for the rate of change of the relative humidity based on the subgrid-scale circulation.

It is appropriate to note that the discussion still concerns subgrid-scale condensation in general, without distinguishing between convective and stratiform clouds. In Eqs. (11) and (12), the same Q appears, but with a different sign, implying that the vapor for the condensational heating is drawn from the immediate vicinity (the same grid box). This view applies well to a stratiform situation; in the case of moist convection, the vertical integral of Eqs. (11) and (12) over the convective layer should be considered. However, the general validity of the statement made in conjunction with Eq. (13) is not altered. Proceeding to a deduction of specific relations, it is natural to regard stratiform and convective condensation separately owing to the fundamental difference in energy conversions taking place in the two cases.

1. Stratiform Condensation

For the stratiform regime, it appears natural that, for condensation to be allowed in a grid box, the relative humidity must exceed a threshold value U_0. Most likely, there is no unique value, but rather U_0 depends on a number of factors. It is obvious that U_0 should increase with decreasing grid distance. (For a sufficiently high resolution, the threshold value is the same as the saturation, implying that the stratiform cloud is resolved by the grid.) All other factors are functions of the current atmospheric state and boundary conditions—for example, wind speed and direction with respect to the subgrid-scale (topography) structure of the earth's surface. The question in this connection is also how high up into the troposphere those conditions have an impact. For low temperatures (< 232 K), it may be reasonable to employ a higher than average threshold value (as alluded to in Section II.B in connection with the discussion of in-cloud conditions in cirrus).

Furthermore, it is common to let the threshold value approach unity at model levels nearest to the ground to suppress an unrealistic readiness of the models to develop clouds in those layers. We close this discussion by observing that the threshold value is the same as the relative humidity of the cloudless part of the grid box at the moment condensation commences therein.

Returning to the closure question, expressed in terms of relative humidity tendency as seen from Eq. (13), we note that the rate of change of U is composed of the rate of change of relative humidity U_0 in the cloud-free part, and of the rate of change of cloud cover b [Eq. (10a)]. To deduce an expression for the U tendency, we write the principal relation for the partitioning mentioned in the first paragraph of this section. The amount of the available vapor, which goes into an increase of the cloud cover and a change of the humidity of the cloudless part, is a fraction H_0 of M_q:

$$\left(q_s - q_0 + \frac{m}{b} \right) \frac{\partial b}{\partial t} + (1 - b) \frac{\partial q_0}{\partial t} = H_0 M_q + E_0 \qquad (14)$$

where $q_0 = U_0 \cdot q_s$ and E_0 is evaporation from precipitation and possible reevaporation of cloud water. A restriction on H_0 is that it must tend to zero as b approaches unity. To obtain a closed set of equations, it is necessary to differentiate Eq. (10a) with respect to time and to adopt an assumption for either how b (or U_0) changes during the condensation process, or how U_0 changes with b. An assumption regarding H_0 is necessary. In Sundqvist et al. (1989) the hypotheses $H_0 = (1 - b)$ and $U_0 = U_{00} + b(U_s - U_{00})$ were adopted; U_{00} is thus the threshold value just at the beginning of condensation in the grid box. Several different formulations of b have been used or suggested in other studies (e.g., Ramanathan et al., 1983; Slingo, 1987; Slingo and Slingo, 1991). Our knowledge is indeed incomplete as to which one of seemingly equally plausible formulations is most useful. Further observations and model experiments should lead to improved insights and more reliable formulations. It is often argued that cloud cover cannot be derived with sufficient precision from relative humidity. At the same time we observe that a study by Kvamstø (1991) indicates that, among several direct and combined quantities, it is just relative humidity that shows the clearest relation to stratiform cloud cover. It is probable that the relative humidity is not sufficiently distinctive to be used alone to get a generally accurate diagnosis of cloud cover, but some other parameter(s) must be considered simultaneously. Even if we do not yet know which is the right, or optimum, approach, it is of prime importance that the actual cloud cover formulation be consistently taken into account in the deduction of the partitioning expression. Rather than discuss some specific cloud-cover formula, the main objective of the above demonstration has been to highlight the basic questions connected with a consistent derivation of a parameterization scheme for subgrid-scale stratiform condensation.

2. Convective Condensation

It is not possible here to give an elaborate treatment for the parameterization of convection. Instead, we will emphasize aspects of the problem that differ from the stratiform case, and we will further illustrate some general questions concerning subgrid-scale parameterization.

The condition for convective condensation to appear is naturally based on a convective instability test, possibly accompanied by some constraint. This test can take different forms, depending on what type of parameterization scheme is adopted. In the case of moist adiabatic adjustment schemes, the condition consists of a stability check and a threshold humidity criterion; if the condition is satisfied, a vertical redistribution is performed regardless of whether there is convergence or divergence of moisture (Manabe and Strickler, 1964). The condition in Kuo-type schemes (Kuo, 1965, 1974) consists of a requirement that there is moisture convergence in the column; basically, the surface air must have positive buoyancy at its lifting condensation level (LCL), since heating in this parameterization is assumed to be proportional to the difference between the temperature along a cloud ascent curve (moist adiabat) through the LCL and the environmental (grid-point) temperature.

The various conditions for convective condensation to appear are generally quite demanding, especially when applied to climate models where the grid distance typically is a few hundred kilometers. It is not obvious that such a large area as a whole has to show a convectively unstable stratification for abundant convection to occur. But it is difficult to deduce a less stringent condition of acceptable validity, as it then would be necessary to involve assumptions about subgrid-scale circulation and characteristics of the earth's surface.

Regarding the closure question, practically all of the various schemes for parameterization of convection that are used in models at present account for only the release of latent heat and neglect the cloud cover. There is still a closure assumption behind those schemes, implying that there is a formulation for the partitioning of the available vapor. This in turn means, as we have seen from the above discussion, that an implicit assumption has been made about cloud cover. An illustrative example of the general partitioning problem is provided by the Kuo (1965) scheme. The sharing of vapor between heating and moistening in that scheme has been of concern for some time (e.g., Kuo, 1974; Anthes, 1977; Krishnamurti et al., 1980; Sundqvist, 1988), because the originally suggested scheme rendered an unrealistic moistening of the convective grid column.

There is a much larger variety of formulations of convective cloud cover than there is of stratiform clouds. In some models, convective cloud cover is neglected, although anvils might be considered (e.g., Harshvardhan et al., 1989). Some models exhibit more elaborate schemes. Slingo (1987) determined the convective cloud cover from an empirical relation with the convective precipitation rate predicted by the model. Sundqvist et al. (1989) resorted to a relation essentially based on the convective intensity and the cloud depth obtained from the model's convection scheme.

G. Additional Viewpoints

In this review, we have emphasized general formulations for the parameterizations of clouds in large-scale models. It is the value, or functional form, of the obtained parameters that account for different types or situations of condensation–cloud processes. Such differentiation has been made in the preceding discussion with regard to both microphysical and subgrid-scale aspects. Still it is relevant to bring up a couple of salient points.

In stratocumulus situations particular features may influence the cloud characteristics. The close relation that exists between stratocumulus structure on one hand and turbulent fluxes in the subcloud layer and cloud-top entrainment features on the other hand, has to be considered in the cloud cover description (Deardorff, 1980; Fravalo et al., 1981; Randall, 1980, 1984a, 1984b; Randall et al., 1985). The interaction between the cloud cover and the radiative fluxes has an important impact on the resulting intensity of the cloud-top entrainment (Randall, 1980; Randall et al., 1985). In this connection we also note a vertical subgrid-scale problem that shows up at cloud top. Due to the radiative flux divergence, there is a substantial cooling effect at cloud top, which penetrates only a fraction of an ordinary model layer. This important effect has to be parameterized without developing numerical instabilities. Cooling effects at cloud top are also important for both convective and cirrus clouds, since the resulting destabilization generally enhances condensation.

As mentioned in the introduction, advection of cloud water is omitted in many, if not most, of the the models that carry this quantity. This may be a serious omission in a global model, especially the advection of cirrus, which can have a strong impact on the formation of cirrus and on the humidity distribution in the upper troposphere. That this may be the case is indicated by the following. The ratio between cirrus cloud-water content (Heymsfield, 1977) and the saturation specific humidity at cirrus levels (200–400 mb) lies in the range 0.1—2 when the saturation humidity value is taken with respect to ice; this wide range is due partly to the large span in cirrus water concentrations and partly to the precise choice of temperature and pressure level. (The corresponding ratio for a liquid water cloud, say nimbostratus, around 700 mb is typically of the order 0.1.) Thus, since the water content of cirrus is approximately the same magnitude as the saturation humidity at the cloud altitude, cirrus clouds can survive quasi-horizontal transportation over very large distances—a moderate evaporative reduction may render a substantial change of the environmental relative humidity. This is a plausible explanation for the far-reaching cirrus veils that originate from the anvils of the convective cloud systems in the tropics.

Our present knowledge regarding subgrid-scale cloud parameterization is so imperfect that we are not able to state what the validity of the various suggested cloud-cover formulations is in relation to model resolution (see, e.g., Kristjánsson, 1991). Therefore, we really do not know if the same cloud-cover algorithm is applicable both in GCM with a resolution of around 200 km and in climate models with resolutions no finer than 500 km. It appears plausible that algorithms for NWPM and GCM are fundamentally different from cloud cover formulas applicable to coarse-resolution climate models. In the latter type of models, it is quite likely that grid boxes will contain areas with both convective and stratiform clouds. Thus, in addition to the questions enumerated above, it will be necessary to set conditions for how convection and stratiform condensation share the grid mesh area.

Another problem is to decide the degree of overlapping for subgrid-scale clouds that appear at different altitudes. Whether the overlapping should be considered random or more systematic is probably very much dependent on the grid resolution and the type of cloud or cloud system (Tian and Curry, 1989). This question is important for radiation calculations. But the extent of the assumed overlapping also has an effect on the conditions for evaporation of precipitation.

III. Data Needs

Observational data are needed for the development of parameterization formulations, as well as for verification of predictions and simulations based on those parameterizations. The ideal situation is to have an ongoing interplay between observational studies and model development. That is, insights gained from observations should be used for the development and improvement of models, the results of which in turn could be used for designing observational strategies. But this is feasible only to a limited extent, since observational programs are complex and costly. However, it is of utmost importance to promote enhanced contacts and cooperation between those working on observational programs and instrumentation, and those involved in model development and simulations. This view of scientific interaction underlies the FIRE (First ISCCP Regional Experiment) program (Cox et al., 1987; Starr, 1987), which has been in operation for several years, and the ARM (Atmospheric Radiation Measurement) initiative (U.S. Department of Energy, 1990), the implementation of which has just begun. There are other equally important international subprograms of the World Climate Research Programme (WCRP) under development. All those programs will be of great value for development of parameterization schemes and for test runs in simplified (e.g., one-dimensional) models. Measurements will also provide the very valuable data needed for testing algorithms used to relate measured radiances to cloud parameters.

For verification of cloudiness in model simulations, it is essential that comparisons between observed and modeled clouds be performed over large areas and long, or at several different, time periods; the ISCCP (International Satellite Cloud Climatology Project) data set provides an important source in this respect (Schiffer and Rossow, 1983; WCRP, 1984). The WCRP has initiated several studies and programs, which are in various stages of development. The Global Precipitation Climatology Project (GPCP of WCRP) should provide very useful precipitation data. Different aspects of observational data are discussed below.

A.   Data for Development of Parameterization Schemes

For the development of cloud parameterization schemes, it is necessary that rele vant data are available on both the microphysical scale and the mesoscale. Intensive observation programs like FIRE and ARM will help to provide such data. Mesoscale data appear more difficult to collect, for two main reasons. First, as implied by the discussions in Sections II.F and II.G, it is not clear what are the most important quantities to measure. Second, it may be necessary to measure subtle features such as differences of temperature, humidity, and vertical wind between cloudy and cloud-free areas, and to measure over scales that are barely resolved, especially with regard to stratiform cloudiness.

For the parameterization of cloud microphysical processes, the following measurements in different synoptic situations and geographic locations have the highest priority (in addition to measurements of conventional meteorological parameters, temperature, humidity, winds, and so on):

1. Three-dimensional distribution of cloud water concentration, distinguishing between liquid and ice phase amounts;
2. Three-dimensional distribution of precipitation, distinguishing between liquid and ice phase amounts.

The second requirement may seem somewhat unreachable; at the time of writing, only the vertically integrated amounts of cloud water and precipitation can be obtained. However, the development of microwave measuring techniques will provide possibilities, in the near future, for obtaining a vertical differentiation of cloud water contents.

When the above information is obtained, it should provide insights into the growth rate of cloud particles and relations between cloud water content and the release of precipitation. It is desirable that rates of precipitation be measured as a function of height. But data from the second item listed above, combined with an assumed terminal fallspeed, will provide information on the rate of evaporation and melting of precipitation.

B.   Data for Verification of Cloud Parameterizations

An important aspect in the development of a cloud parameterization scheme is its verification. So far, there have been very few specific studies of simulated cloudiness in different numerical models; also, there is a serious void of suitable data for such verification. Today, there are powerful potentials for improvements in this respect. For example, cloud cover derived from satellite radiance data (Arking and Childs, 1985; Stowe et al., 1985) were used by Slingo (1987) to verify the simulated cloud amounts of the ECMWF (European Center for Medium Range Weather Forcasts) global model.

Ground-based cloud observations are useful, but they have limitations—first because they are essentially confined to land areas, and second because they can be made only during daylight. But a careful and scrutinizing analysis of this type of observation yields a significant empirical data set, complementary to ISCCP data, for verification of climate model simulations (Hughes, 1984).

The ERBE (Earth Radiation Budget Experiment) data set, used in the concept of cloud radiative forcing (Ramanathan et al., 1989), is also an important means of verification of model clouds, primarily of their integral properties.

It is necessary that data on liquid water content as well as cloud cover become available for verification. Adequate amounts of data on liquid water content can be obtained only from satellite measurements. Raustein et al. (1991) compared simulated cloud cover and cloud water content with the corresponding quantities interpreted from satellite measurements and projected on the same grid as the model was run on. Cloud cover was derived from AVHRR (Advanced Very High Resolution Radiometer) data and water content from both AVHRR and SSM/I (Special Sensor Microwave Imager) data. The comparison shows that these types of retrieved quantities constitute a potentially powerful means for quantitative verification of simulations. The simulations were performed with a mesoscale model (Sundqvist et al., 1989) with a horizontal resolution of 50 km. Results from the paper by Raustein et al. are exhibited in Figs. 2–5. A few algorithms have been suggested for extracting cloudiness from radiance data (Curry et al., 1990; Saunders, 1989; Takeda and Liu, 1987). To achieve the best possible accuracy and validity in these interpretations, they should be calibrated against direct measurements.

In discussions on use of satellite data for model verification, the view is often put forth that one should compare the model's outgoing radiation at the top of the atmosphere with satellite measurements. An example of such an approach may be found in Pudykiewicz et al. (1992). The direct use of observed radiances would eliminate introduction of uncertainties that follow if interpretation algorithms are employed to transform the observations into cloud parameters. However, it seems impossible to avoid the need for inversions to cloudiness quantities. For example, assuming that differences are found between simulated and measured radiances, then, presuming that clouds strongly influence the characteristics of the outgoing radiation, it becomes necessary to trace the differences in terms of observed and simulated clouds to find deficiencies in the cloud parameterization. Consequently, it is essential that interpretation algorithms be developed. Extensive testing and

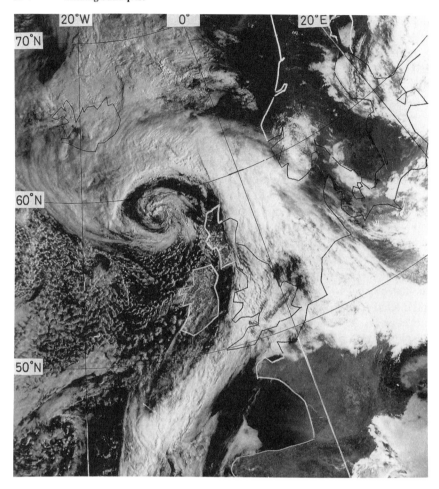

Figure 2   Satellite (NOAA 10) picture in near infrared (channel 2) from passage 27 August 1988 around 0845 UTC. (Satellite Receiving Station, Dundee University, U.K. From Raustein et al., 1991.)

use of those schemes will involve continuous improvement, implying a reduction of uncertainty in the retrieved cloudiness.

C.   Initial Data for Prediction and Simulation

The inclusion of cloud water content as a dependent variable in a model system implies that an initial state of the cloud water is required for the integration. The demand for accuracy of this cloud water analysis is naturally most pronounced for

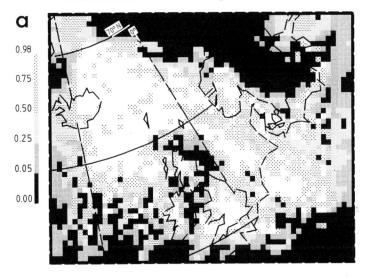

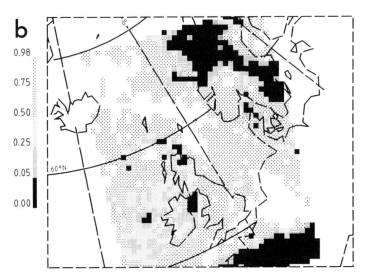

Figure 3   Picture of two-dimensional cloud cover. (a) From 21-h model prediction, valid time 0900 UTC 27 August 1988. (b) From interpretation of AVHRR data of the same time as in Fig. 2. The scale to the left shows the shading for the different fractional cover intervals. The retrieved data cover only part of the model domain. (From Raustein et al., 1991.)

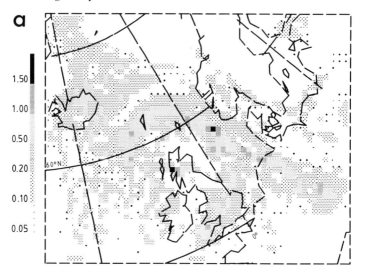

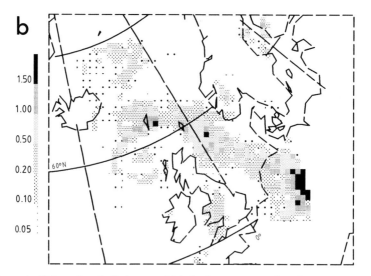

Figure 4   Picture of vertically integrated (a) cloud water content from 21-h model prediction, valid time 0900 UTC 27 August 1988, and (b) liquid water amount from interpretation of AVHRR data of the same time as in Fig. 2. The scale to the left shows the shading for the different amount intervals; unit, kg m^{-2} (or mm water column). The retrieved data cover only part of the model domain. (From Raustein et al., 1991.)

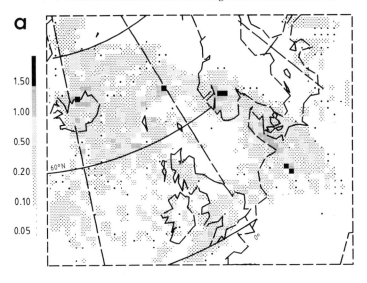

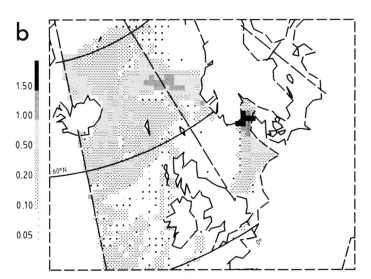

Figure 5   Picture of vertically integrated (a) cloud water content from 30-h of the model prediction, valid time 1800 UTC 27 August 1988, and (b) liquid water amount from interpretation of SSM/I data from the period 17.50 UTC (eastern part) 27 August 1988 to 1930 UTC (western part) 27 August 1988. (The retrieval from SSM/I is possible only over sea.) The scale to the left shows the shading for the different amount intervals; unit, kg m^{-2} (or mm water column). The retrieved data cover only part of the model domain. (From Raustein et al., 1991.)

prediction models. The analysis of cloudiness cannot be performed as an individual task; rather, it has to be carried out in conjunction with a humidity analysis. To achieve the required accuracy for a meaningful cloud water analysis, it is necessary that the humidity analysis be carried out with a refinement that is considerably better than is customary today. Assuming there is a realistic initial cloud field available, this would be nearly wiped out during the first few time-steps of the integration if the initial moisture field is the generally smooth humidity analysis that characterizes present-day models. To improve this situation, the humidity data from both radiosondes and satellite measurements have to be analyzed while various constraints, given by the flow field, are taken into account so that regions with high humidity, in particular, are maintained during the analysis procedure. Various attempts to improve the initial state for numerical integration are made by inclusion of diabatic effects in connection with nonlinear normal mode initialization, which deals with an adjustment between the wind and mass fields (e.g., Heckley et al., 1990). The purpose of this inclusion is to obtain an improved divergence field, which then has a favorable feedback on the diabatic processes. However, such attempts are probably futile, unless the humidity analysis is improved. The importance of this is indicated both by experiments with physical initialization (Krishnamurti et al., 1984, 1991) and in the attempts with cumulus initialization (Donner, 1988; Donner and Rasch 1989). In his study of diabatic initialization and modification to the moisture field, Turpeinen (1990) states "that in spite of the diabatic initialization, the spin-up times remain long (6–9 hours) if no humidity enhancement is applied."

Furthermore, it is interesting to note that dynamic initialization (used by Krishnamurti et al., 1984, 1991) and initialization by digital filtering (Lynch and Huang, 1992) not only suppress unrealistic high-frequency pressure oscillations, but also render redistributions of the humidity field toward a physically improved agreement with the flow field. The latter effect follows because these techniques are based on time integration of the model equations around the initial time. This means that in a model system that contains cloudiness equations, the initialization process can produce initial fields of cloud water and cloud cover. These may be used directly as initial fields, but ideally they should be used as first guesses in a more comprehensive analysis of atmospheric water (vapor, liquid, solid), based on observational data.

The data obtained by remote sensing (radiances) so far gives only vertically integrated cloud water, precipitation, and cloud cover. (There are promising indications that some measurements will yield some degree of vertical resolution.) Hence, to obtain vertical resolution, it is necessary to resort to some assumption or model (Tao et al., 1990). The best approach in this respect ought to be to use the model's cloud parameterization scheme, possibly assuming steady state. This approach would have a positive feedback effect, in that additional emphasis would be put on the need for a reliable, widely valid cloud parameterization.

IV. Concluding Remarks

In this review, we have considered basic questions involved in the derivation of a consistent cloud parameterization scheme, which allows for the prediction of hydrometeors and fractional cloud cover. The formulation of this problem is divided into two main parts: parameterization of the microphysical processes (i.e., mechanisms for release of precipitation and for evaporation and melting of precipitation) and parameterization of mesoscale features (i.e., cloud cover and the amount of released latent heat).

The question of parameterization of mesoscale processes is quite different from the question of microscale parameterization. We do possess some knowledge of the latter, albeit incomplete, that helps us derive formulations that have some realism and reliability. However, our knowledge of the parameterization of mesoscale features is regrettably poor. In this respect we have to resort to hypotheses and empiricism. To acquire satisfactory insights into this problem area, substantial research efforts must be devoted to both theoretical and model studies, and to suitable observation/measurement programs. Closer cooperation between theorists, field researchers, and modelers in this area should be encouraged and promoted.

Questions regarding data needs have also been addressed. Dedicated observation programs are essential for the gathering of data sufficiently detailed for parameterization development, and for the calibration of retrieval algorithms used for the conversion of satellite radiances into cloudiness parameters. The detailed measurements and observations may be obtained by a combination of different means, such as ground-based sets of instruments and suitably equipped aircraft. The latter also provide a powerful platform for qualitative observations of mesoscale circulations.

For verification and validation of model predictions and simulations, data of regional and global scales are needed. It seems that adequate amounts of such data can be provided only through satellite measurements. Data from ISCCP are of this type and are especially useful for climate modeling contexts. For the purpose of a more precise verification, it is also necessary to have access to (near) instantaneous data, not only of cloud cover but also of cloud water content and precipitation. It is furthermore stressed that this category of observational data is needed for the creation of initial cloudiness fields, and for vital additional information on humidity, a parameter of vital importance in cloud models.

Acknowledgments

The author wishes to thank an anonymous reviewer and Professor Peter V. Hobbs for constructive suggestions and advice on important references, and for remarks on a couple of formal errors in equations. This work was supported in part by the Swedish Natural Science Research Council (NFR) under grants Ö-ER 2923–301 and E-EG 2923–302.

References

Anthes, R. A. (1977). A cumulus parameterization scheme utilizing a one-dimensional cloud model. *Mon. Wea. Rev.* **105**, 270–286.

Arking, A. and Childs, J. D. (1985). Retrieval of cloud cover parameters from multispectral satellite images. *J. Clim. and Appl . Meteor.* **24**, 322–333.

Berge, E. (1990). A regional numerical sulfur dispersion model using a meteorological model with explicit treatment of clouds. *Tellus* **42B**, 389–407.

Bergeron, T. (1935). On the physics of clouds and precipitation. *Proc. 5th Assembly UGGI* Lisbon, 2.

Browning, K. A. (1985). Conceptual models of precipitation systems. *The Meteorological Magazine* **114**, 293–319.

Browning, K. A. (1990). Organization of clouds and precipitation in extratropical cyclones. *Extratropical Cyclenes. The Erik Palmén Memorial Volume*, Chester W. Newton and Eero O. Holopainen (eds.), American Meteor. Soc., Boston, Mass. 129–153.

Cess, R. D., Potter, G. L., Blanchet, J. P., Boer, G. J., Gahn, S. J., Kiehl, J. T., Le Treut, H., Liang, Z.-X., Mitchell, J. F. B., Morcrette, J-J., Randall, D. A., Riches, M. R., Röckner, E., Schlese, U., Slingo, A., Taylor, K. E., Washington, W. M., Weatherald, R. T., and Yagai, I. (1989). Interpretation of cloud–climate feedback as produced by 14 atmospheric general circulation models. *Science* **245**, 513–516.

Charlson, R. J., Lovelock, J. E., Andreae, M. O., and Warren, S. G. (1987). Oceanic phytoplancton, atmospheric sulfur, cloud albedo, and climate. *Nature* **326**, 655–661.

Cotton, William R., and Anthes, Richard A. (1989). *Storm and Cloud Dynamics.* Academic Press, New York.

Cox, S. K., McDougal, D., Randall, D. A., and Schiffer, R. A. (1987). FIRE—The first ISCCP regional experiment. *Atmosphere–Ocean* **67**, 114–118.

Curry, J. A., Ardeel, C. D., and Tian, L. (1990). Liquid water content and precipitation characteristics of stratiform clouds as inferred from satellite microwave measurements. *J. Geophys. Res.* **95**, 16659–16671.

Deardorff, J. W. (1980). Stratocumulus-capped mixed layers derived from a three-dimensional model. *Bdy. Layer Meteor.* **7**, 495–527.

Donner, Leo J. (1988). An initialization for cumulus convection in numerical weather prediction models. *Mon. Wea. Rev.* **116**, 377–385.

Donner, L. J., and Rasch, P. J. (1989). Cumulus initialization in a global model for numerical weather prediction. *Mon. Wea. Rev.* **117**, 2654–2671.

Findeisen, W. (1938). Die kolloid meteorologischen Vorgänge bei der Niederschlagung. *Meteor. Z.* **55**, 121–131.

Flatøy, F. (1992). Comparison of two parameterization schemes for cloud and precipitation processes. *Tellus* **44A**, 41–53.

Fouquart, Y., Buriez, J. C., Herman, M., and Kandel, R. S. (1990). The influence of clouds on radiation: a climate-modeling perspective. *Reviews of Geophysics* **28**, 145–166.

Fravalo, C., Fouquart, Y., and Rosset, R. (1981). The sensitivity of a model of low stratiform clouds to radiation. *J. Atmos. Sci.* **38**, 1049–1062.

Golding, B. W. (1986). Short-range forecasting over the United Kingdom using a mesoscale forecasting system. Short- and medium-range numerical weather prediction. pp. 563–572 in *Collection of Papers Presented at the WMO/IUGG NWP Symposium*. Tokyo, 4–6 August 1986. (T. Matsuno, ed.), Meteorological Society of Japan.

Harshvardhan, Randall, D. A., Corstti, T. G., and Dazlich, D. A., 1989. Earth radiation budget and cloudiness simulations with a general circulation model. *J. Atmos. Sci.* **46**, 1922–1942.

Heckley, W. A., Kelly, G., and Tiedtke, M. (1990). Sensitivity of ECMWF analyses-forecasts of tropical cyclones to cumulus parameterization. *Mon. Wea. Rev.* **118**, 1743–1757.

Hense, A., and Heise, E. (1984). A sensitivity study of cloud parameterization in general circulation models. *Beiträge zur Physik der Atmosphäre* **57**, 240–258.

Heymsfield, A. J. (1977). Precipitation development in stratoform ice clouds: A microphysical and dynamical study. *J. Atmos. Sci.* **34**, 367–381.

Heymsfield, A. J., and Sabin, R. M. (1989). Cirrus crystal nucleation by homogeneous freezing of solution droplets. *J. Atmos. Sci.* **46**, 2252–2264.

Heymsfield, A. J. and Donner, L. J., 1990. A scheme for parameterizing ice-cloud water content in general circulation models. *J. Atmos. Sci.* **47**, 1865–1877.

Hughes, N. A. 1984. Global cloud climatologies: A historical review. *J. Climate and Appl. Meteor.* **23**, 724–751.

IPCC (1990). Climate change. The IPCC scientific assessment. Report prepared for Intergovernmental Panel on Climate Change by Working group I. WMO/UNEP. Cambridge University Press.

Kessler, E. (1969). On the distribution and continuity of water substance in atmospheric circulation. *Meteor. Monog.* **10**, American Meteor. Soc., Boston, Mass.

Krishnamurti, T. N., Ramanathan, Y., Pan, Hua-Lu, Pasch, R. J., and Molinari, J. (1980). Cumulus parameterization and rainfall rates. *Mon. Wea. Rev.* **108**, 456–464.

Krishnamurti, T. N., Ingles, K., Cocke, S., Kitade, T., and Pasch, R. (1984). Details of low latitude medium-range numerical weather prediction using a global spectral model. Part II. Effects of orography and physical initialization. *J. Meteor. Soc. Japan* **62**, 613–649.

Krishnamurti, T. N., Xue, J., Bedi, H. S., Ingles, K., and Oosterhof, D. (1991). Physical initialization for numerical weather prediction over the tropics. *Tellus* **43A-B**, 53–81.

Kristjánsson, J. E. (1991). Cloud parameterization at different horizontal resolutions. *Quart. J. Roy. Meteor. Soc.* **117**, 1255–1280.

Kuo, H. L. (1965). On formation and intensification of tropical cyclones through latent heat release by cumulus convection. *J. Atmos. Sci.* **22**, 40–63.

Kuo, H. L. (1974). Further studies of the parameterization of the influence of cumulus convection on large-scale flow. *J. Atmos. Sci.* **31**, 1232–1240.

Kvamstø, Nils Gunnar (1991). An investigation of relations between stratiform fractional cloud cover and other meteorological parameters in numerical weather prediction models. *J. Appl. Meteor.* **30**, 200–216.

Lelieveld, J., and Crutzen, P. J. (1991). The role of clouds in tropospheric photochemistry. *J. Atmos. Chemisty* **12**, 229–267.

Locatelli, J. D., and Hobbs, P. V. (1987). The mesoscale and microscale structure and organization of clouds and precipitation in mid-latitude cyclones. XIII: Structure of warm front. *J. Atmos. Sci.* **44**, 2290–2309.

Lynch, P., and Huang, X.-Y. (1992). Initialization of the HIRLAM model using a digital filter. *Mon. Wea. Rev.* **120**, 1019–1034.

Manabe, S., and Strickler, R. F. (1964). Thermal equilibrium of the atmosphere with convective adjustment. *J. Atmos. Sci.* **21**, 361–385.

Marshall, J. S. and Palmer W. McK. (1948). The distribution of raindrops with size. *J. of Meteor.* **5**, 165–166.

Mason, B. J. (1971). *The Physics of Clouds.* Oxford University Press.

Matveev, L. T. (1984). *Cloud Dynamics.* D. Reidel Publishing Company.

Mitchell, J. F. B., Senior, C. A., and Ingram, W. J. (1989). CO_2 and climate: a missing feedback? *Nature* **341**, 132–134.

Pruppacher, H. R., and Klett, J. D. (1978). *Microphysics of Clouds and Precipitation.* D. Reidel Publishing Company.

Pudykiewicz, J., Benoit, R., and Mailhot, J. (1992). Inclusion and verification of a predictive cloud water scheme in a regional numerical weather prediction model. *Mon. Wea. Rev.* **120**, 612–626.

Ramanathan, V., and Collins W. (1991). Thermodynamic regulation of ocean warming by cirrus clouds deduced from observations of the 1987 El Niño. *Nature* **351**, 27–32.

Ramanathan, V. E., Pitcher, E. J., Malone, R. C., and Blackmon, M. L. (1983). The response of a spectral general circulation model to refinements in radiative processes. *J. Atmos. Sci.* **40**, 605–630.

Ramanathan, V., Cess, R. D., Harrison, E. F., Minnis, P., Barkstrom, B. R., Ahmad, E., Hartmann, D. (1989). Cloud-radiative forcing and climate: Results from the Earth Radiation Budget Experiment. *Science* **243**, 57–63.

Randall, D. A. (1980). Entrainment into a stratocumulus layer with distributed radiative cooling. *J. Atmos. Sci.* **37**, 148–159.

Randall, D. A. (1984*a*). Buoyant production and consumption of turbulence kinetic energy in cloud-topped mixed layers. *J. Atmos. Sci.* **41**, 402–413.

Randall, D. A. (1984*b*). Stratocumulus cloud deepening through entrainment. *Tellus* **36A**, 446–457.

Randall, D. A., Abeles, J. A., and Corsetti, T. G. (1985). Seasonal simulations of the planetary boundary layer and boundary-layer stratocumulus clouds with a general circulation model. *J. Atmos. Sci.* **42**, 641–676.

Randall, D. A., Harshvardhan, Dazlich, D. A., and Corsetti T. G. (1989). Interaction among radiation, convection, and large-scale dynamics in a general circulation model. *J. Atmos. Sci.* **46**, 1943–1970.

Rangno, A. L., and Hobbs, P. V. (1991). Ice particle concentrations and precipitation development in small polar maritime cumuliform clouds. *Quart. J. Roy. Meteor. Soc.* **117**, 207–241.

Raustein, E., Sundqvist, H., and Katsaros, K. B. (1991). Quantitative comparison between simulated cloudiness and clouds objectively derived from satellite data. *Tellus* **43A**, 306–320.

Röckner, E., and Schlese, U. (1987). Cloud optical depth feedbacks and climate modeling. *Nature* **329**, 138–140.

Rutledge, S. A., and Hobbs, P. V. (1983). The mesoscale and microscale structure and organization of clouds and precipitation in mid-latitude cyclones VIII: A model for the "seeder–feeder" process in warm-frontal rainbands. *J. Atmos. Sci.* **40**, 1185–1206.

Sasamori, T. (1975). A statistical model for stationary atmospheric cloudiness, liquid water content and rate of precipitation. *Mon. Wea. Rev.* **12**, 1037–1049.

Saunders, R. W. (1989). A comparison of satellite-retrieved parameters with mesoscale model analyses. *Quart. J. Roy. Meteor. Soc.* **115**, 651–672.

Schiffer, R. A., and Rossow, W. B. (1983). The International Cloud Climatology Project (ISCCP): The first project of the World Climate Research Programme. *Atmos.–Ocean.* **64**, 779–784.

Slingo, J. M. (1987). The development and verification of a cloud prediction scheme for the ECMWF model. *Quart. J. Roy. Meteor. Soc.* **113**, 899–927.

Slingo, A. (1990). Sensitivity in the earth's radiation budget to changes in low clouds. *Nature* **343**, 49–51.

Slingo, J. M., and Slingo, A. (1991). The response of a general circulation model to cloud longwave radiation forcing. II: Further studies. *Quart. J. Roy. Meteor. Soc.* **117**, 333–364.

Smith, R. N. B. (1990). A scheme for predicting layer clouds and their water content in a general circulation model. *Quart. J. Roy. Meteor. Soc.* **116**, 435–460.

Sommeria, G., and Deardorff, J. W. (1977). Subgrid-scale condensation in models of nonprecipitating clouds. *J. Atmos. Sci.* **34**, 344–355.

Starr, D. O'C., and Cox, S. K. (1985). Cirrus clouds, Part I: A cirrus cloud model. *J. Atmos. Sci.* **42**, 2663–2681.

Starr, D. O'C. (1987). A cirrus-cloud experiment: Intensive field observations planned for FIRE. *Atmos.–Ocean.* **68**, 119–124.

Stephens, G. L., and Webster, P. J. (1981). Clouds and climate: Sensitivity of simple systems. *J. Atmos. Sci.* **38**, 235–247.

Stowe, L. L., Pellegrino, P. P., Hwang, P. H., Bhartia, P. K., Eck, T. F., Wellemeyer, C. G., Read, S. M., and Long, C. S. (1985). Use of Nimbus-7 satellite data for verification of GCM-generated cloud cover. Report from Workshop on Cloud Cover Parameterization in Numerical Models, 26–28 November 1984, ECMWF, Reading, UK.

Sundqvist, H. (1978). A parameterization scheme for nonconvective condensation including prediction of cloud water content. *Quart. J. Roy. Meteor. Soc.* **104**, 677–690.

Sundqvist, H. (1988). Parameterization of condensation and associated clouds in models for weather prediction and general circulation simulation. In *Physically-based Modelling and Simulation of Climate and Climatic change*, M. E. Schlesinger (ed.), Reidel Pub. Co. Part 1, 433–461.

Sundqvist, H., Berge, E., and Kristjánsson, J. E. (1989): Condensation and cloud parameterization studies with a mesoscale numerical weather prediction model. *Mon. Wea. Rev.* **117**, 1641–1657.

Takeda, T., and Liu, G. (1987). Estimation of atmospheric liquid-water amount by NIMBUS 7 SMMR data: A new method and its application to the western North-Pacific region. *J. Meteor. Soc. Japan* **65**, 931–947.

Tao, W-K., Simpson, J., Lang, S., McCumber, M., Adler, R., and Penc, R. (1990). An algorithm to estimate the heating budget from vertical hydrometer profiles. *J. Appl. Meteor.* **29**, 1232–1244.

Tian, Lin, and Curry, Judith A. (1989). Cloud overlap statistics. *J. Geophys. Res.* **94**, D-7, 9925–9935.

Turpeinen, Olli M. (1990). Diabatic initialization of the Canadian regional finite-element (RFE) model using satellite data. Part II: Sensitivity to humidity enhancement, latent-heating and rain rates. *Mon. Wea. Rev.* **118**, 1396–1407.

U.S. Department of Energy (1990). Atmospheric Radiation Measurement Program Plan. DOE/ER-0441, February 1990. U.S. Department of Energy, Office of Energy Research, Washington, D.C. 20585.

WCRP (1984). International Satellite Cloud Climatology Project (ISCCP). Report from the Third Session of the International Working Group on Data Management (Tokyo, 6–8 March 1984). WCP—82. ICSU/WMO.

Wetherald, R. J., and Manabe, S. (1988). Cloud feedback processes in a general circulation model. *J. Atmos. Sci.* **45**, 1397–1415.

Chapter 8   Stratospheric Aerosols and Clouds

M. Patrick McCormick
Atmospheric Sciences Division
NASA Langley Research Center
Hampton, Virginia

Pi-Huan Wang
Science and Technology Corporation
Hampton, Virginia

Lamont R. Poole
Atmospheric Sciences Division
NASA Langley Research Center
Hampton, Virginia

This chapter summarizes recent advances in our understanding of stratospheric aerosols and polar stratospheric clouds (PSCs), with an emphasis on satellite-based observations from the SAM II (Stratospheric Aerosol Measurement II), SAGE I (Stratospheric Aerosol and Gas Experiment I), and SAGE II sensors. Stratospheric aerosols and PSCs influence the earth's radiation balance by interacting with solar and terrestrial radiation. They also catalyze heterogeneous chemical reactions that can markedly alter stratospheric odd nitrogen, chlorine, and ozone levels. Seasonal variations in aerosol levels are described, and the effects of recent volcanic eruptions that have significantly perturbed the stratosphere are presented. Of special interest is the June 1991 eruption of Mount Pinatubo, which appears to have been the largest of the century. The Pinatubo aerosols caused an approximately 3°C increase in the daily zonal mean stratospheric temperature at low latitudes by the northern fall of 1991, and may lead to a globally averaged surface cooling of about 0.5°C by late 1992. Finally, areas for improvement in future stratospheric aerosol measurements are recommended.

I. Introduction

The primary radiative importance of stratospheric aerosols lies in the fact that they interact directly with solar and terrestrial radiation (e.g., Hansen et al., 1990). By scattering solar radiation back to space, the aerosols have a cooling effect at the earth's surface, whereas by absorbing upwelling infrared radiation, they induce warming in the stratospheric layer(s) where the particles reside. Stratospheric warmings caused by major volcanic events have been reported by Newell (1970) for the 1963 Mount Agung volcanic eruption, by Labitzke et al. (1983) and Fujita (1985) for the 1982 El Chichon eruption, and by Labitzke and McCormick (1992) for the 1991 Mount Pinatubo eruption. Cooling of the earth's surface following major volcanic eruptions has been suggested in many historical papers and reported by Hansen and Lebedeff (1988), Angell and Korshover (1983), and Robock (1991).

Stratospheric aerosols also may influence the earth's radiative balance indirectly through their role as cloud condensation nuclei (CCN). For example, they are involved in the formation of polar stratospheric clouds (PSCs) (e.g., Poole and McCormick, 1988) and possibly in the large-scale development of cirrus clouds (Mohnen, 1990). The number and properties of CCN have been shown to affect stratus cloud particle concentration, size, and optical properties (Twomey, 1977a; Twomey et al., 1984; Coakley et al., 1988). It is possible that stratospheric aerosols transported to the upper troposphere have a similar effect on the properties of cirrus and, hence, their interactions with solar and terrestrial radiation.

Stratospheric aerosols and clouds also play important roles in stratospheric chemistry. Recent investigations (Hofmann and Solomon, 1989; Rodriguez et al., 1991; Mather and Brune, 1990; Brasseur et al., 1991; Prather, 1992) have suggested that stratospheric aerosols catalyze heterogeneous chemical reactions that alter odd nitrogen and chlorine levels and, hence, may affect stratospheric ozone levels. Similar but much more efficient reaction sequences on PSC particles have been shown to be crucial in the formation of the Antarctic ozone hole (Solomon, 1990; Poole et al., 1992). Because stratospheric ozone also plays an important role in establishing the temperature structure of the stratosphere (e.g., Brasseur and Solomon, 1984) and radiative forcing of the surface–troposphere system (Ramaswamy et al., 1992), changes in stratospheric aerosols may further modify the radiation budget through their effects on global ozone levels.

It should also be noted that the performance of remote sensors designed to measure various atmospheric parameters can be affected either directly or indirectly by aerosols. As an example, the Laser Atmospheric Wind Sounder (LAWS), which is intended to measure winds in the upper troposphere (Baker and Curran, 1985), relies on backscattering from aerosols as its signal source. On the other hand, during periods of high aerosol loading, such as those following major volcanic eruptions, it is imperative to correct the remotely sensed signals of key atmospheric constituents such as H_2O or O_3 for aerosol contamination (Bandeen and Fraser, 1982; DeLuisi et al., 1989; Fleig et al., 1990).

There have been substantial advances in our understanding of stratospheric aerosols and clouds since the discovery of the stratospheric aerosol layer some three decades ago by Junge et al. (1961). Good general reviews of aerosol chemistry and physics can be found in Craig (1965), Twomey (1977b), Whitten (1982), Deepak (1982), Gerber and Deepak (1984), and Hobbs and McCormick (1988). Reviews of the state of knowledge of the more recently discovered PSCs can be found in Solomon et al. (1990) and Poole et al. (1992). A development of particular note during the past decade has been comprehensive, near-global remote sensing observations of stratospheric particulates. These observations have been invaluable in tracking and quantifying large disturbances to the stratosphere caused by the eruptions of El Chichon in 1982 and Mount Pinatubo in 1991. This chapter will summarize recent advances in stratospheric aerosol and PSC research, with an emphasis on satellite-based observations and the insights gained therefrom on the atmospheric effects of volcanic eruptions.

II. General Characteristics of Stratospheric Aerosols and Clouds

A complete description of stratospheric aerosols would require data on their composition, refractive index, size distribution, and shape—a suite of information rarely, if ever, available. There is sufficient evidence, however, that the bulk of ambient stratospheric aerosols can be represented reasonably well by spherical liquid droplets whose composition by weight at normal stratospheric temperatures is about 75% H_2SO_4 and 25% H_2O, which implies, in turn, a refractive index of about 1.42 at visible wavelengths (Rosen, 1971). It also appears that the size distribution of nonvolcanic aerosol particles is roughly lognormal (Pinnick et al., 1976; Russell et al., 1981). Immediately after a volcanic eruption, however, the average particle size increases and the size distribution becomes more complex, perhaps multimodal (Oberbeck et al., 1983).

The sources, sinks, and distribution of stratospheric aerosols have been reviewed by Turco et al. (1982) and Mohnen (1990). Perhaps the most important chemical pathway by which these aerosols are formed is oxidation of SO_2 in the stratosphere and subsequent formation of H_2SO_4 molecules, followed by nucleation of H_2SO_4/H_2O droplets and their growth by condensation.

Advection, sedimentation, evaporation, and coagulation are all important to the distribution of stratospheric aerosols. Details of the droplet nucleation process are still uncertain (Mohnen, 1990), but it is generally believed that the tropical stratosphere is the source region and the aerosols produced there are transported to higher latitudes through large-scale circulation processes. The most significant sources of stratospheric aerosols are major volcanic eruptions that can inject large amounts of SO_2 into the stratosphere (see Section IV). Crutzen (1976) suggested that the persistence of the aerosol layer during prolonged periods without major volcanic eruptions is due to the diffusion into the stratosphere of carbonyl sulfide (OCS) originating from biogenic or anthropogenic sources in the troposphere. Recently, Hofmann (1990) reported that there had been a 5% annual increase in background (nonvolcanic) stratospheric aerosol mass over the period from 1978 to 1989. Similar decadal increases in stratospheric aerosol optical depth and integrated backscatter have been noted (Yue and Poole, 1991; Poole et al., 1992). Hofmann (1991) further speculated that the secular increase in aerosol mass is closely related to the increase in sulfur emissions from aircraft flying in the stratosphere and upper troposphere.

Polar stratospheric clouds were first identified and reported by McCormick et al. (1982) from recurrent high-extinction anomalies in the SAM II data record during polar winters. PSCs are thought to exist in two primary categories: Type I, composed of nitric acid–water particles that are stable at temperatures above the frost point; and Type II, composed of water ice crystals that are stable at sub-frost-point temperatures. There are considerable uncertainties in the process(es) by which PSCs are nucleated (Hofmann et al., 1990; Schlager et al., 1990; Dye et al., 1992), as well as in the precise composition of Type I clouds (Tolbert and Middle-brook, 1990; Ritzhaupt and Devlin, 1991; Smith et al., 1991), which were previously thought to be solely nitric acid trihydrate (Hanson and Mauersberger, 1988). Laboratory measurements (e.g., Molina et al. 1987; Tolbert et al., 1987; Leu, 1988; Hanson and Ravishankara, 1991) imply that both PSC types are efficient sites for heterogeneous reactions that activate chlorine radicals from normally benign reservoirs and, at the same time, sequester odd nitrogen species as less reactive HNO_3. These reactions "prime" the polar stratosphere for chlorine-catalyzed ozone depletion, a process that can be quite rapid and severe if the PSC particles involved grow large enough to undergo appreciable sedimentation and irreversibly remove the sequestered odd nitrogen.

III. Long-Term Observations by Satellite and Ground-Based Lidar

Near-global monitoring of stratospheric aerosols has been accomplished by satellite instruments (Table 1), complemented with ground-based lidar and *in situ* measurements using various instruments on balloon or aircraft platforms (Pues-

Table 1

Satellite Limb Extinction Measurements

Experiment	Satellite	Launch	Latitude coverage	Wavelength (μm) (primary species)
SAM II (solar)	NIMBUS-7	Oct. 1978[a]	64°–80°N 64°–80°S	1.0 (aerosol)
SAGE I (solar)	AEM-2	Feb. 1979[b]	79°N–79°S	0.385 (NO_2) 0.450 (aerosol) 0.600 (O_3) 1.0 (aerosol)
SAGE II (solar)	ERBS	Oct. 1984[a]	80°N–80°S	0.385 (aerosol) 0.448 (NO_2) 0.453 (aerosol) 0.525 (aerosol) 0.600 (O_3) 0.940 (H_2O) 1.0 (aerosol)

[a] Presently operational.
[b] Obtained data through November 1981.

chel, 1991). The SAM II sensor was launched aboard Nimbus 7 in October 1978, SAGE I aboard the Application Explorer Mission (AEM) 2 spacecraft in February 1979, and SAGE II aboard the Earth Radiation Budget Satellite (ERBS) in October 1984. The SAM II and SAGE II instruments are still operational at the time of this writing. All of these instruments measure, during spacecraft sunrise and sunset, solar radiation as it traverses the limb of the earth's atmosphere as a function of tangent height. By referencing the transmitted radiation to the measured exoatmospheric solar radiation, the extinction due to aerosols as a function of altitude is readily determined. This solar occultation technique is thus self-calibrating (McCormick, 1987), allowing for the accurate determination of long-term trends.

The SAM II sensor provides single-wavelength (1.0 μm) aerosol extinction measurements which, due to the sun-synchronous Nimbus 7 orbit, are all located in the two polar regions. SAGE I was placed in an orbit tailored to complement the geographical coverage of SAM II and provide mid- and low-latitude aerosol extinction data at wavelengths of 1.0 μm and 0.45 μm. SAGE II is in an orbit similar to that of SAGE I and measures aerosol extinction at four wavelengths (1.02, 0.525, 0.453, and 0.385 μm).

A. SAM II Observations

Figure 1 shows long-term records of the weekly averaged SAM II 1.0-μm optical depth (the integral of the aerosol extinction coefficient from the tropopause +2 km upward through 30 km) for the Arctic and Antarctic regions. The 1978–79 period is generally referred to as the aerosol "background" state since it followed a period of about 5 years without major volcanic activities and preceded the series of eruptions in the early 1980s (Table 2). The dates of various volcanic eruptions thought to have injected material into the stratosphere are marked along the abscissa in Fig. 1. The April 1982 El Chichon eruption produced the largest perturbation observed during the 1980s, increasing the optical depth in the Arctic by a factor of 40 over the 1979 background value of about 1.5×10^{-3}.

Evident in Fig. 1 are prominent enhancements in optical depth during local winter (and early Antarctic spring) that are signatures of PSC formation. Another prominent seasonal feature is the optical depth minimum that occurs during each Antarctic spring, following the PSC events. Between 1979 and 1981, the value of the minimum optical depth dropped below the 10^{-3} level. Similar, but shallower, minima are seen in the Arctic record during spring. The optical minimum is thought to be the result of both the sedimentation of large PSC particles and subsidence of the cold winter air in the polar region vortex (Wang and McCormick, 1985*a;* Kent et al., 1985; Thomason and Poole, 1992). Analysis also indicates that increases in zonal mean aerosol optical depth in the polar region during stratospheric warming periods are related to the poleward transport of aerosols by planetary waves (Wang and McCormick, 1985*b*).

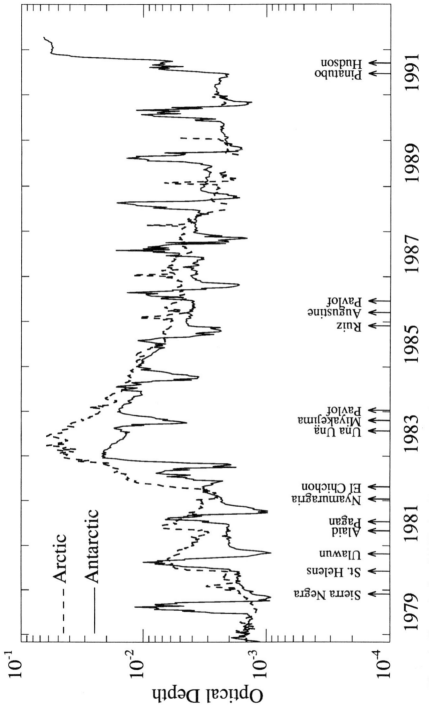

Figure 1 Time series from SAM II of weekly averaged values of aerosol optical depth (at a wavelength of 1.0 μm) at high latitudes (integrated from 2 km above the tropopause upward).

Table 2

Estimate of Aerosol Mass Loading in the Stratosphere

Date	Volcano or background	Location	Mass loading (in units of 10^5 tons)	Source
1979	Background		5.7	Kent and McCormick (1984)
1989	Background		7.5	Yue et al. (1992)
March 17, 1963	Agung	8.4°S, 115.5°E	160	Deirmendjian (1973)
			300	Cadle et al. (1976, 1977)
October 10, 1974	Fuego	14.5°N, 90.9°E	60	Cadle et al. (1976, 1977)
			30	Lazrus et al. (1979)
January 22, 1976	St. Augustine	59.4°N, 153.4°W	6.0	Cadle et al. (1977)
April 17, 1979	La Soufriere	13.3°N, 61.2°W	0.023	McCormick et al. (1981)
November 13, 1979	Sierra Negra	0.8°S, 91.2°W	1.6	Kent and McCormick (1984)
May 18, 1980	St. Helens	46.2°N, 122.2°W	5.5	Kent and McCormick (1984)
October 7, 1980	Ulawun	5.0°S, 151.3°E	1.8	Kent and McCormick (1984)
April 27, 1981	Alaid	50.8°N, 155.5°E	3.0	Kent and McCormick (1984)
May 15, 1981	Pagan	18.1°N, 145.8°E	2.0	Kent and McCormick (1984)
April 4, 1982	El Chichon	17.3°N, 93.2°W	200	Hofmann and Rosen (1983)
			120	McCormick and Swissler (1983)
			101	Mroz et al. (1983)
November 13, 1985	Ruiz	4.9°N, 75.4°W	5.6	Yue et al. (1992)
February 10, 1990	Kelut	7.9°S, 112.3°E	1.8	Yue et al. (1992)
June 15, 1991	Pinatubo	15.1°N, 120.4°E	≈300–400	McCormick and Veiga (1992)

B.   SAGE I and II Aerosol Observations

To study seasonal variations in the spatial distribution of stratospheric aerosols, SAGE II data can be used to derive seasonally averaged zonal mean profiles of 1.02-μm aerosol extinction ratio (the ratio of aerosol extinction coefficient to molecular extinction coefficient). Plate 5 shows the results for 1989 northern hemisphere spring (March–May), summer (June–August), fall (September–November), and winter (December 1989–February 1990). These seasonal distributions show the stratospheric aerosol layer clearly, with its center at an altitude of about 22 km in the tropics, indicative of a tropical aerosol source region. Very similar results were produced from SAGE I observations. In terms of typical aerosol size, which can be estimated from the multiwavelength data, smaller particles are generally found in the tropical lower stratosphere, and larger particles are found at higher altitudes above the tropical tropopause and at higher latitudes (Yue and Deepak, 1984). This feature is indicative of the growth of particles through microphysical processes as they are transported to regions with lower temperatures.

Outside the tropics, the center of the layer slopes with latitude, but it is located at a roughly constant differential altitude above the local tropopause ($+$ symbols in Plate 6 indicate the local tropopause height). The maxima located near the tropical tropopause indicate high clouds in that region. Note the difference in aerosol distribution between the two hemispheres. This feature may be a consequence of hemispheric differences in planetary wave activity, shown to be responsible for similar differences in mean zonal winds, temperature, and total ozone (Geller and Wu, 1987). Note also the gaps in the aerosol distribution near the subtropical tropopause, especially during the local winter season. These are indicative of the troposphere–stratosphere exchange associated with the subtropical jet stream (Shapiro and Keyser, 1990). A recent study using SAGE I and SAGE II data shows evidence of intrusion of stratospheric aerosols into the upper troposphere at most latitudes, except in the tropics and possibly at very high latitudes (Kent et al., 1991). Further investigation is necessary for a full understanding of this exchange process.

To study long-term trends in stratospheric aerosols, SAGE I and II aerosol data can be used to derive the time history of hemispherically averaged optical depth (Fig. 2). There were at least six volcanic eruptions during the SAGE I operating period, from February 1979 to November 1981. These eruptions resulted in a nearly steady, gradual increase of aerosol optical depth in the southern hemisphere. In the northern hemisphere, the optical depth changes were more dramatic and a much higher maximum value (about 3.2×10^{-3}) was reached some 4 months after the May 1980 eruption of Mount St. Helens. By late 1981, the globally averaged aerosol optical depth was about a factor of 2 higher than the "background" level observed in early 1979.

SAGE II observations (Fig. 2b) show large values of aerosol optical depth in

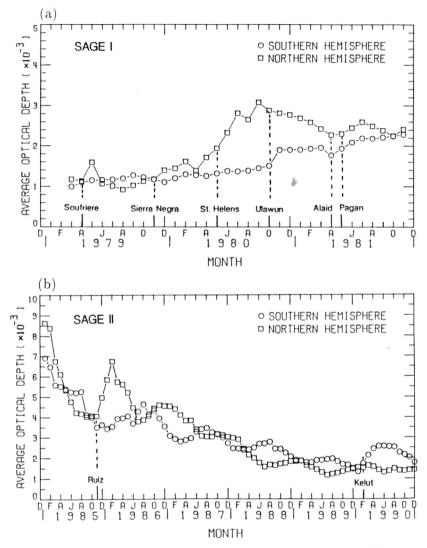

Figure 2   Time series of hemispherically averaged aerosol optical depth: (a) SAGE I observations ($\lambda = 1.0$ μm); and (b) SAGE II observations ($\lambda = 1.02$ μ).

early 1985, a residual effect of the April 1982 El Chichon eruption. Late in the record shown in Fig. 2b, some 8 years after the El Chichon eruption, the stratospheric aerosol loading reaches values near the background conditions of early 1979. A seasonal variation is also very evident (Yue et al., 1992), with maxima in local winter. The impact of the November 1985 Ruiz eruption can easily be seen

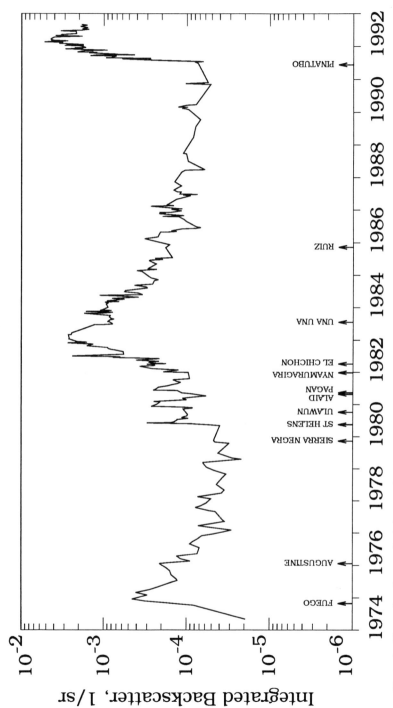

Figure 3 Time series of stratospheric integrated aerosol backscatter at 694 nm wavelength measured by the 48-inch lidar system at the NASA Langley Research Center (37°N, 76°W).

in the northern hemisphere record, and the effect of the February 1990 Kelut eruption can be seen in the southern hemisphere record. The figure shows that the peak
in aerosol loading can be expected from 1 to 4 months following an eruption, a
delay thought to be related to the time required for gas-to-particle conversion of
the SO_2 emitted by the volcano.

C.   Lidar Observations at the NASA Langley Research Center

Figure 3 shows the long-term record (1974–1991) of integrated lidar backscatter
at a wavelength of 694 nm obtained at the NASA Langley Research Center (37°N,
76°W). The times of major volcanic eruptions are noted as in previous figures.
There was a magnitude enhancement of approximately an order of 2 following the
El Chichon eruption, and a larger enhancement due to the Mount Pinatubo eruption is evident. Figure 4 shows a more detailed comparative record of the effects
of Pinatubo and El Chichon. About 100 days after the respective eruptions, the
integrated backscatter values resulting from the two eruptions were similar over
Langley. Between then and about 300 days, the integrated backscatter due to
Pinatubo exceeded that experienced after El Chichon by 50–100%, after which
the two records approached about the same value. A more nearly global perspective is given below.

IV.   Observations from the Eruption of Mount Pinatubo

The Philippine volcano Mount Pinatubo (15.1°N, 120.4°E) erupted violently in
mid-June 1991, emitting massive plumes that penetrated well into the strato-

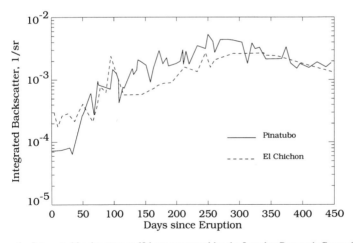

Figure 4   Integrated backscatter at 694 nm measured by the Langley Research Center lidar following the eruptions of El Chichon and Mount Pinatubo.

sphere, reaching altitudes as high as 40 km. Using TOMS (Total Ozone Mapping Spectrometer) data, Bluth et al. (1992) estimated that 20 Mt of SO_2 were injected into the stratosphere from the eruption. This is more than three times that produced by the 1982 El Chichon eruption (Krueger, 1983). A record 40 Mt of H_2SO_4/H_2O aerosols from Mount Pinatubo are expected. A series of papers in the January 1992 issue of *Geophysical Research Letters* gives a preliminary assessment of the stratospheric and climatic effects of the Pinatubo eruption (McCormick, 1992).

The satellite observations from TOMS (Bluth et al., 1992) and NOAA-11 AVHRR (Stowe et al., 1992) showed that the Pinatubo plume first encircled the earth in about 3 weeks. SAGE II aerosol extinction data show that by about 45 days after the eruption, the volcanic material at altitudes above 20 km had spread between 10°S and 30°N. Below 20 km, the material had spread to higher northern latitudes (McCormick and Veiga, 1992). By late August and September 1991, the aerosol material above 20 km had dispersed throughout much of the Southern Hemisphere and the southern edge of the dense tropical belt of material had moved to about 20°S. At the same time, the densest layer of material above 20 km in the Northern Hemisphere had spread to about 30°–35°N. These SAGE II results are corroborated by measurements from ground-based lidar systems (Jäger, 1992; Post et al., 1992), airborne lidar (Winker and Osborn, 1992), and AVHRR (Stowe et al., 1992).

To illustrate the impact of the Pinatubo eruption, SAGE II sunrise aerosol extinction profile measurements at 1.02 μm during the period July 10–20, 1991 (Plate 6) are compared with the corresponding results from the previous year (Plate 7). The upper panels indicate the locations (latitude as dots, longitude as solid squares) of the SAGE II measurements. The regions color-coded as black are cases where the aerosol extinction exceeds the sensitivity of the instrument. In 1990, such cases were due to obscuration of the sun by thick clouds in the troposphere. SAGE II first encountered aerosols from the Pinatubo eruption aerosols on July 11, 1991, near a latitude of 10°S. Between July 11 and 19, the SAGE II measurements reached below 20 km in only a few cases because of the heavy tropical aerosol loading at that time. The Pinatubo cloud appeared densest between 10°S and 30°N, and most aerosols were located at altitudes below 28 km. Values of aerosol optical depth higher than 0.4 at mid-visible wavelengths were measured in the tropics at this time (Valero and Pilewskie, 1992). Inhomogeneity in the spatial distribution of the Pinatubo cloud is illustrated by the wave structure between 22 and 28 km.

To illustrate the evolution of the Pinatubo aerosols, Plate 8 presents a series of four longitude–latitude distributions of 1.02-μm aerosol optical depth constructed from SAGE II observations. Plate 8a depicts the situation about a month before the June 15 eruption, in which the green color corresponds to background (ambient) values of optical depth. About one month following the eruption, material from Pinatubo had spread in a belt about the equator, as shown in Plate 8b by the increasingly higher levels of optical depth. Mid-latitude anticyclones in both hemispheres transported material to high latitudes as shown in this figure. In the north, transport occurred preferentially in the upper troposphere and lower stratosphere near 16 km and was associated with the Asian monsoon. In the south, transport occurred primarily between 18 and 24 km and was caused by a large-scale westwardly propagating high-pressure system. Plate 8c shows the optical depth distribution about 3 months after the eruption. The dense tropical belt had become more uniform and had spread to higher latitudes, especially northward to about 30°N. Slightly less dense material had spread to high southern latitudes, and the still less dense material had spread to high northern latitudes, caused at least in part by the anticyclonic systems mentioned above. Plate 8d depicts the situation about 7 months after the eruption. Note that the densest material had by then covered most of the northern hemisphere and had reached high southern latitudes also.

It is believed that the conversion of the volcanic SO_2 to sulfuric acid vapor and subsequently to sulfuric acid particles is relatively rapid with an e-folding time on the order of about a month (Turco, 1991). An airborne mission staged in the Caribbean in July 1991 (McCormick, 1992) included a combination of instruments to investigate this gas-to-particle conversion in the fresh volcanic plume at a time when the conversion was most active. Results from the correlation spectrometer and the Fourier transform spectrometer consistently showed an SO_2 column amount of approximately 10^{16} molecules cm^{-2} (Hoff, 1992; Mankin et al., 1992). This result is comparable to the earlier and higher value determined from the TOMS instrument considering an e-folding time of about a month, and agrees also with the high-resolution ground-based infrared observations conducted at Mauna Loa Observatory (Goldman et al., 1992). It is also important to note that rapid gas-to-particle conversion processes are evident in the increase in the magnitude of the backscatter ratio and the height at which the maximum ratio occurred as observed by the Mauna Loa lidar system (DeFoor et al., 1992; see also McCormick and Veiga, 1992). The lag time between the eruption and the occurrence of peak aerosol mass, however, appears to be about 4 months as mentioned previously, but additional data are needed to confirm this finding.

Another important data set needed to determine the impact of the Pinatubo eruption is the composition and size distribution of the volcanic aerosols. The University of Wyoming/NOAA balloon results indicate that more than 95% of particles were composed of an H_2SO_4/H_2O solution in all layers sampled in July 1991, when a portion of the volcanic cloud appeared over Laramie, Wyoming. In addition, a bimodal size distribution was observed. Hardly any of the sulfate particles analyzed showed evidence of a solid or dissolved nucleus. Hence, the observed high concentrations of Pinatubo aerosols suggest homogeneous or ion nucleation as the most possible aerosol production mechanism (Deshler et al., 1992; Sheridan et al., 1992).

Possible signs of the effect of volcanic aerosols on stratospheric minor species has been reported following the Pinatubo eruption. Column amounts of NO_2 over Lauder, New Zealand were smaller by more than 40% in October 1991 compared to the previous year (Johnston et al., 1992). As noted by these authors, these observations may provide supporting evidence for the notion mentioned earlier that heterogeneous processes can convert N_2O_5 to HNO_3 on the surface of sulfate aerosols. If this proves to be the case, volcanically enhanced aerosols may accelerate global ozone depletion in the presence of high chlorine levels.

As mentioned in the introduction, stratospheric aerosols cause a cooling at the earth's surface and a local warming at altitudes where they are present. Analyses of September and October 1991 temperature data indicate that the monthly averaged zonal mean 30-mb temperatures at 20°N were approximately 2.5°C above the 26-year mean, and some daily zonal mean positive anomalies of 3°C occurred (Labitzke and McCormick, 1992). The change in the global earth surface temperature is more difficult to assess. However, by using a global climate model, Hansen et al. (1992) predicted that a noticeable aerosol-induced global average surface cooling would begin in late 1991, with the largest effect occurring in late 1992 depending on the timing of any El Niño. Because of the enormous quantity of Pinatubo aerosols, Hansen and colleagues state that this volcanic episode should provide a strict test of the accuracy of current climate models, which are predicting a global mean surface cooling that is three standard deviations from the mean value.

V. Concluding Remarks

Long-term observations from satellite remote sensors, complemented by lidar and *in situ* measurements, have significantly improved our understanding of the global climatology of stratospheric aerosols over the last decade. Clearly, volcanically produced SO_2 is the major source of stratospheric aerosols. Possible increases in background (nonvolcanic) aerosol concentrations, due to increased sulfur emissions by high-altitude aircraft, have been suggested (Hofmann, 1990). Two massive volcanic eruptions have occurred during the last 10 years, El Chichon (1982) and Mount Pinatubo (1991). Perturbations of stratospheric aerosol loadings from these eruptions reached at least two orders of magnitude above background levels. Based on TOMS SO_2 measurements and SAGE II aerosol measurements, it appears that Pinatubo produced two to three times the aerosol generated by El Chichon, making it the largest of the century in terms of its impact on stratospheric aerosols. About a 3°C increase in the daily zonal mean stratospheric temperature at low latitudes has been reported (Labitzke and McCormick, 1992). Meanwhile, a measurable global average surface cooling resulting from the Pinatubo event is expected to be observed in 1992 (Hansen et al., 1992). In view of their climatic importance and the difficulties in predicting volcanic eruptions, a continuous monitoring of the global behavior of stratospheric aerosols is essential. Also, careful measurements of ozone and gases important to the stratospheric ozone budget should be closely monitored for any possible effects of aerosol-catalyzed heterogeneous chemistry.

Although there have been significant recent advances in stratospheric aerosol research, many areas remain in which improvements are needed. For example, only the column amount of SO_2 injected into the stratosphere by a volcanic eruption can be measured by the existing TOMS satellite instrument. No information on SO_2 vertical distributions is available, nor is the TOMS technique sensitive to low levels of SO_2. With respect to aerosols, size distribution data are essential. Also, because of possible high loadings of volcanic aerosols, the dynamic range and sensitivity of satellite-based aerosol extinction measurements needs improvement so that saturation conditions can be avoided in the future. The design of the SAGE III sensor scheduled to fly as part of the Earth Observing System (EOS) should mitigate these problems with a greatly increased dynamic range and an expanded and differential wavelength measurement capability.

Acknowledgments

The authors are grateful to G. K. Yue, R. E. Veiga, and K. Skeens for their helpful assistance in preparation of the manuscript. P.-H. Wang is supported by NASA contract NAS1–18676.

References

Angell, J. K., and Korshover, J. (1983). *Mon. Wea. Rev.* **111**, 2129–2135.
Baker, W. E., and Curran, R. J. (eds.) (1985). "Report of the NASA Workshop on Global Wind Measurements," 70 pp. A. Deepak Publishing, Hampton, Virginia.
Bandeen, W. R., and Fraser, R. S. (1982). *NASA TM-84959*, National Aeronautics and Space Administration, Greenbelt, Maryland.
Bluth, G., Doiron S. D., Schnetzler, C. C., Krueger, A., and Walter, L. S. (1992). *Geophys. Res. Lett.* **19**, 151–154.
Brasseur, G., and Solomon, S. (1984). "Aeronomy of the Middle Atmosphere," 441 pp. D. Reidel Publishing Company.
Brasseur, G. P., Granier, C., and Walters, S. (1991). *Nature* **348**, 626.
Cadle, R. D., Kiang, C. S., and Louis, J. F. (1976). *J. Geophys. Res.* **81**, 3125–3132.
Cadle, R. D., Fernald, F. G., and Frush, C. L. (1977). *J. Geophys. Res.* **82**, 1783–1786.
Coakley, J. A., Jr., Berstein, R. L., and Durkee, P. A. (1988). "Aerosols and Climate" (Hobbs and McCormick, eds.), 253–260. A. Deepak Publishing, Hampton, Virginia.
Craig, R. A. (1965). "The Upper Atmosphere, Meteorology, and Physics," 509 pp. Academic Press, New York.
Crutzen, P. J. (1976). *Geophys. Res. Lett.* **3**, 73–76.
Deepak, A. (ed.) (1982). "Atmospheric Aerosols, Their Formation, Optical Properties and Effects," 480 pp. Spectrum Press, Hampton, Virginia.
DeFoor, T. E., Robinson, E., and Ryan, S. (1992). *Geophys. Res. Lett.* **19**, 187–190.
Deirmendjian, D. (1973). *Advan. Geophys.* **16**, 267–296.

DeLuisi, J. J., Longenecker, D. U., Mateer, C. L., and Wuebbles, D. J. (1989). *J. Geophys. Res.* **94**, 9837–9846.

Deshler, T., Hofmann, D. J., Johnson, B. J., and Rozier, W. B. R. (1992). *Geophys. Res. Lett.* **19**, 199–222.

Dye, J. E., Baumgardner, D., Gandrud, B. W., Kawa, S. R., Kelly, K. K., Loewenstein, M., Ferry, G. V., Chan, K. R., and Gary, B. L. (1992). *J. Geophys. Res.* **97**, 8015–8034.

Fleig, A. J., McPaters, R. D., Bhartia, P. K., Schlesinger, B. M., Cebula, R. P., Klenk, K. F., Taylor, S. L., and Heath, D. F. (1990). *NASA RP 1234*, NASA Goddard Space Flight Center, Greenbelt, Maryland.

Fujita, T. (1985). *Meteor. and Geophys.* **36**, 47–60.

Geller, M. A., and Wu, M-F (1987). "Transport Processes in the Middle Atmosphere" (G. Visconti and R. Garcia, eds.), 3–17. D. Reidel Publishing, Boston, Massachusetts.

Gerber, H. E., and Deepak, A. (eds.) (1984). "Aerosols and Their Climate Effects," 297 pp. A. Deepak Publishing, Hampton, Virginia.

Goldman, A., Murcray, F. J., Rinsland, C. P., Blatherwick, R. D., David, S. J., Murcray, F. H., and Murcray, D. G. (1992). *Geophys. Res. Lett.* **19**, 183–186.

Hansen, J. E., and Lebedeff, S. (1988). *Geophys. Res. Lett.* **15**, 323–326.

Hansen, J. E., Rossow, W., and Fung, I. (1990). *Issues in Science and Technology* **Fall 1990**, 62–69.

Hansen, J., Lacis, A., Ruedy, R., and Sato, M. (1992). *Geophys. Res. Lett.* **19**, 215–218.

Hanson, D., and Mauersberger, K. (1988). *Geophys. Res. Lett.* **15**, 855–858.

Hanson, D. R., and Ravishankara, A. R. (1991). *J. Geophys. Res.* **96**, 5081–5090.

Hobbs, P., and McCormick, M. P., (eds.) (1988). "Aerosols and Climate," 486 pp. A. Deepak Publishing, Hampton, Virginia.

Hoff, R. (1992). *Geophys. Res. Lett.* **19**, 175–178.

Hofmann, D. J. (1990). *Science* **248**, 996-1000.

Hofmann, D. J. (1991). *Nature* **349**, 659.

Hofmann, D. J., and Rosen, J. M. (1983). *Geophys. Res. Lett.* **10**, 313–316.

Hofmann, D. J., Deshler, T., Arnold, F., and Schlager, H. (1990). *Geophys. Res. Lett.* **17**, 1279–1282.

Hofmann, D. J., and Solomon, S. (1989). *J. Geophys. Res.* **94**, 5029–5041.

Jäger, H. (1992). *Geophys. Res. Lett.* **19**, 191–194.

Johnston, P. V., McKenzie, R. L., Keys, J. G., and Matthews, W. A. (1992). *Geophys. Res. Lett.* **19**, 211–213.

Junge, C. E., Chagnon, C. W., and Manson, J. E. (1961). *J. Meteor.* **18**, 81–108.

Kent, G., and McCormick, M. P. (1984). *J. Geophys. Res.* **89**, 5303–5314.

Kent, G., Trepte, C. R., Farrukh, U. O., and McCormick, M. P. (1985). *J. Atmos. Sci.* **42**, 1536–1551.

Kent, G. S., McCormick, M. P., and Schaffner, S. K. (1991). *J. Geophys. Res.* **96**, 5249–5267.

Krueger, A. J. (1983). *Science* **220**, 1377–1378.

Labitzke, K., and McCormick, M. P. (1992). *Geophys. Res. Lett.* **19**, 207–210.

Labitzke, K., Naujokat, B., and McCormick, M. P. (1983). *Geophys. Res. Lett.* **10**, 24–26.

Lazrus, A. L., Cadle, R. D., Gandrud, B. W., Greenberg, J. P., Huebert, B. J., and Rose, W. I. Jr. (1979). *J. Geophys. Res.* **84**, 7869–7875.

Leu, M-T (1988). *Geophys. Res. Lett.* **15**, 17–20.

Mankin, W. G., Coffey, M. T., and Goldman, A. (1992). *Geophys. Res. Lett.* **19**, 179–182.

Mather, J. H., and Brune, W. H. (1990). *Geophys. Res. Lett.* **17**, 1283–1286.

McCormick, M. P., Kent, G. S., Yue, G. K., and Cunnold, D. M. (1981). "SAGE Measurements of the Stratospheric Aerosol Dispersion and Loading from the Soufriere Volcano." *NASA Tech. Paper 1922*, November 1981.

McCormick, M. P., Steele, H. M., Hamill, P., Chu, W. P., and Swissler, T. J. (1982). *J. Atmos. Sci.* **39**, 1387–1397.

McCormick, M. P., and Swissler, T. J. (1983). *Geophys. Res. Lett.* **10**, 877–880.

McCormick, M. P. (1987). *Adv. Space Res.* **7**, 319–326.

McCormick, M. P. (1992). *Geophys. Res. Lett.* **10**, 149.
McCormick, M. P., and Veiga, R. E. (1992). *Geophys. Res. Lett.* **19**, 155–158.
Mohnen, V. A. (1990). *J. Atmos. Sci.* **47**, 1933–1948.
Molina, M. J., Tso, T-L, Molina, L. T., and Yang, F. C.-Y. (1987). *Science* **238**, 1253–1257.
Mroz, E. J., Mason, A. S., and Sedlacek, W. A. (1983). *Geophys. Res. Lett.* **10**, 873–876.
Newell, R. E. (1970). *J. Atmos. Sci.* **27**, 977–978.
Oberbeck, V. R., Danielsen, E. F., Snetsinger, K. G., and Ferry, G. V. (1983). *Geophys. Res. Lett.* **10**, 1021–1024.
Pinnick, R. G., Rosen, J. M., and Hofmann, D. J. (1976). *J. Atmos. Sci.* **33**, 304–314.
Poole, L. R., and McCormick, M. P. (1988). *J. Geophys. Res.* **93**, 8423–8430.
Poole, L. R., Jones, R. L., Kurylo, M. J., and Wahner, A. (1992). "Heterogeneous Process: Laboratory, Field, and Modeling Studies." *Global Ozone Research and Monitoring Project Report No. 25.*
Post, M. J., Grund, C. J., and Langford, A. O. (1992). *Geophys. Res. Lett.* **19**, 195–198.
Prather, M. (1992). *J. Geophys. Res.* **97**, 10187–10191.
Pueschel, R. (ed.) (1991). "International Workshop on Stratospheric Aerosols: Measurements, Properties, and Effects." *NASA Conference Paper 3114*, March 27–30, 1990.
Ramaswamy, V., Schwarzkopf, M. D., and Shine, K. P. (1992). *Nature* (in press).
Ritzhaupt, G., and Devlin, J. P. (1991). *J. Chem. Phys.* **90**, 90–95.
Robock, A. (1991). "Greenhouse-Gas-Induced Climate Change," (M. E. Schlesinger, ed.), 429–443. Elsevier, New York.
Rodriguez, J. M., Ko, M. K.W., and Sze, N. D. (1991). *Nature* **352**, 134–137.
Rosen, J. M. (1971). *J. Appl. Meteor.* **10**, 1044–1046.
Russell, P. B., Swissler, T. J., McCormick, M. P., Chu, W. P., Livingston, J. M., and Pepin, T. J. (1981). *J. Atmos. Sci.* **38**, 1279–1294.
Schlager, H., Arnold, F., Hofmann, D., and Deshler, T. (1990). *Geophys. Res. Lett.* **17**, 1275–1278.
Shapiro, M. A., and Keyser, D. (1990). "Fronts, Jet Streams, and the Tropopause," *NOAA Technical Memorandum ERL WLP-182*, National Oceanic and Atmospheric Administration, Wave Propagation Laboratory, Boulder, Colorado.
Sheridan, P. J., Schnell, R. C., Hofmann, D. J., and Deshler, T. (1992). *Geophys. Res. Lett.* **19**, 203–206.
Smith, R. H., Leu, M.-T., and Keyser, L. F. (1991). *J. Phys. Chem.* **95**, 5924–5930.
Solomon, S. (1990). *Nature* **347**, 347–354.
Solomon, S., Grose, W. L., Jones, R. L., McCormick, M. P., Molina, M. J., O'Neill, A., Poole, L. R., and Shine, K. P. (1990). *Global Ozone Research and Monitoring Project Report No. 20.*
Stowe, L. L., Carcey, R. M., and Pellegrino, P. P. (1992). *Geophys. Res. Lett.* **19**, 159–162.
Thomason, L. W., and Poole, L. R. (1992). "Analysis of Antarctic Stratospheric Aerosol Properties Using SAGE II Extinction Measurements," 47–51. Proceedings of the Eighth Conference on the Middle Atmosphere, American Meteorological Society.
Tolbert, M. A., and Middlebrook, A. M. (1990). *J. Geophys. Res.* **95**, 22423–22431.
Tolbert, M. A., Rossi, M. J., Malhotra, R., and Golden, D. M. (1987). *Science* **238**, 1258–1260.
Turco, R. (1991). "Volcanic Aerosols: Chemistry, Microphysics, Evolution, and Effects. Volcanism–Climate Interaction" (L. S. Walter and S. de Silva, eds.), *NASA CP-10062*, NASA, Washington, D.C.
Turco, R. P., Whitten, R. C., and Toon, O. B. (1982). *Rev. Geophys. and Space Phys.* **20**, 233–279.
Twomey, S. A. (1977*a*). *J. Atmos. Sci.* **34**, 1149–1152.
Twomey, S. A. (1977*b*). "Atmospheric Aerosols," 302 pp. Elsevier Scientific Publishing Company, New York.
Twomey, S. A., Piepgrass, M., and Wolfe, T. L. (1984). *Tellus* **36B**, 356–366.
Valero, F. P. J., and Pilewskie, P. (1992). *Geophys. Res. Lett.* **19**, 163–166.
Wang, P.-H., and McCormick, M. P. (1985*a*). *J. Geophys. Res.* **90B**, 10597–10606.
Wang, P.-H., and McCormick, M. P. (1985*b*). *J. Geophys. Res.* **90**, 2360–2364.

Whitten, R. C. (ed.) (1982). "The Stratospheric Aerosol Layer," 152 pp. Springer-Verlag, Berlin Heidelger.

Winker, D. M., and Osborn, M. T. (1992). *Geophys. Res. Lett.* **19**, 167–170.

Yue, G. K., and Deepak, A. (1984). *Geophys. Res. Lett.* **11**, 999–1002.

Yue, G. K., and Poole, L. R. (1991). "Comparison of the Impact of Volcanic Eruptions and Aircraft Emissions on the Aerosol Mass Loading and Sulfur Budget in the Stratosphere." Paper presented at the First Annual High-Speed Research Workshop, Williamsburg, Virginia, May 14–16, 1991.

Yue, G. K., Poole, L. R., and McCormick, M. P. (1992). "Mass Loading of Stratospheric Aerosols from Eruptions of Ruiz and Kelut," (manuscript in preparation).

Index

International Geophysics Series

EDITED BY

RENATA DMOWSKA
Division of Applied Sciences
Harvard University
Cambridge, Massachusetts

JAMES R. HOLTON
Department of Atmospheric Sciences
University of Washington
Seattle, Washington

*Out of print.